INDUSTRIAL DESIGN TECHNIQUES AND MATERIALS

RAYMOND GUIDOT

INDUSTRIAL DESIGN TECHNIQUES AND MATERIALS

RAYMOND GUIDOT

Flammarion

CONTENTS

A BRIEF HISTORY

RAYMOND GUIDOT

DISTANT PAST AND NEW MATERIALS

Design is often considered to stand at the crossroads of the arts. The role of the designer has been described as being to ensure coherence between the technical requirements of manufacture, the internal structure of the object, its use value, and its appearance. The designer therefore requires a significant body of knowledge, one that covers all aspects of the design process: the phases of conception and the various tools involved, the choice of materials and their processes of implementation. Undoubtedly, engineers and technicians are expected to know this, too, but their knowledge tends to be more specialized; their expertise is most often employed to thwart the aims of the competition. The designer more generally intervenes on behalf of clients and his knowledge can be of a less specialized nature.

With their coordinating role, designers have to look on each problem from a fresh angle. Ever-vigilant, they have to be as aware of the advent of new methods, materials, and manufacturing processes as of the development of existing methods and materials. More broadly, designers need a wide-ranging cultural awareness that links form and technique. They have to be aware of those crucial historical junctures—some innovative cultural phenomenon, a significant development, etc.—that interact with technological progress. Their insatiable thirst for innovation may lead them to study the most ancient traditional techniques and objects, sometimes dating to the prehistoric period. Creative genius, focusing on the harmonious unification of form and function in a utilitarian object, is not a modern phenomenon: it was already present some 250,000 years ago in the development of the double-sided hand axe, a distant ancestor of today's multipurpose pocketknife.

The more recent "contemporary" period, which historians date from the end of the eighteenth century when the Industrial Revolution in the United Kingdom entered its active phase, is peppered with discoveries and innovations that had a decisive effect on the production of utility objects. For example, the introduction of coke (a fuel whose mechanical resistance during combustion makes it an advantageous substitute for charcoal) allowed cast-iron to be produced in vast quantities from the very start of the 18th century, which in turn made it possible to achieve such giant feats of engineering as the bridges at Coalbrookdale (1779) and Sunderland (1796).

preceding page
DETAIL OF THE YVETTE LAMP
BY FRANÇOIS AZAMBOURG.

facing page
DETAIL OF COALBROOKDALE
BRIDGE OVER THE SEVERN,
SHROPSHIRE, ENGLAND, 1777–79.

SUNDERLAND BRIDGE, ENGLAND, 1796, BUILT BY ROWLAND BURDON ACCORDING TO A PLAN BY THOMAS PAINE. THE STRUCTURE'S SIX ARCHES ARE SPANNED BY SECTIONS OF CAST-IRON.

COALBROOKDALE BRIDGE OVER
THE SEVERN, SHROPSHIRE,
ENGLAND, 1777–79, TO A PLAN
BY THE ENGINEER JOHN
WILKINSON AND THE ARCHITECT
THOMAS FARNOLLS PRITCHARD. THE
IRON SPANS OF THE CONSTRUCTION
WERE CAST IN THE WORKSHOPS
OF ABRAHAM DARBY III.

following pages
TAKING DOWN THE OLD CAST-IRON
STREET LAMPS IN PARIS,
AROUND 1920–30.

EUR
EUR
OTO
LA COCA

It also allowed items to be molded in large quantities, leading to the mass production of street furniture such as public benches, grills, street lamps, etc., as well as everyday items such as containers for preparing foodstuffs, wood and coal stoves, seating, etc.

Such new varieties of manufacture simultaneously generate new modes of construction. Thus a chair made of cast-iron elements is not mounted with traditional mortise and tenon joints but with screws. To a significant extent, it was these new practices that, around the middle of the nineteenth century, sparked off the debate that pitted the supporters of craftwork against the advocates of industrial production. For the former, copying forms created by expert craftsmen in centuries past, in a noble material such as wood, was preferable to working with cheap materials and poor craftsmanship.

The second half of the nineteenth century saw the development of semi-synthetic materials—chemically improved natural resins such as ebonite, celluloid, galalithe, etc. Again, they were used as, and therefore considered to be, substitutes for more costly materials. Ignoring their intrinsic qualities, they were made to mimic ebony, horn, and ivory. In German, the word for these materials was *ersatz*, which, in French and English, soon became a pejorative term. Nonetheless, in 1909, Bakelite, the first totally synthetic material, one that could only be created in the chemistry lab, was well received. Durable, relatively resistant to chemical agents and high temperatures, capable of being beautifully molded, with virtually scratch-proof surfaces, it bravely paraded its natural, brownish hues enriched by surface effects. An unexpected result of the manufacturing process resulted in different marbling effects in pieces produced in the same mold. This, in turn, introduced the idea of differentiation in mass production. In 1927, this element of chance was deliberately used in the manufacture of the famous Philips "saucer" radio speaker known as the 2007. For certain displays, the operator was instructed to throw into the mold containing the raw material a handful of brass filings that then glittered on the piece's surface, like a random-generated allusion to the Milky Way. These ideas are very close to the research that, from the early 1970s on, underpinned the work of a designer such as Gaetano Pesce.

FERDINAND KRAMER,
CAST-IRON COAL-BURNING STOVE, GERMANY, 1926.

The modernity of phenolic resin was confirmed when it was chosen to mold cases or bodies for those appliances emerging from the applied sciences (record player, cameras, radio sets) that originated in the nineteenth century

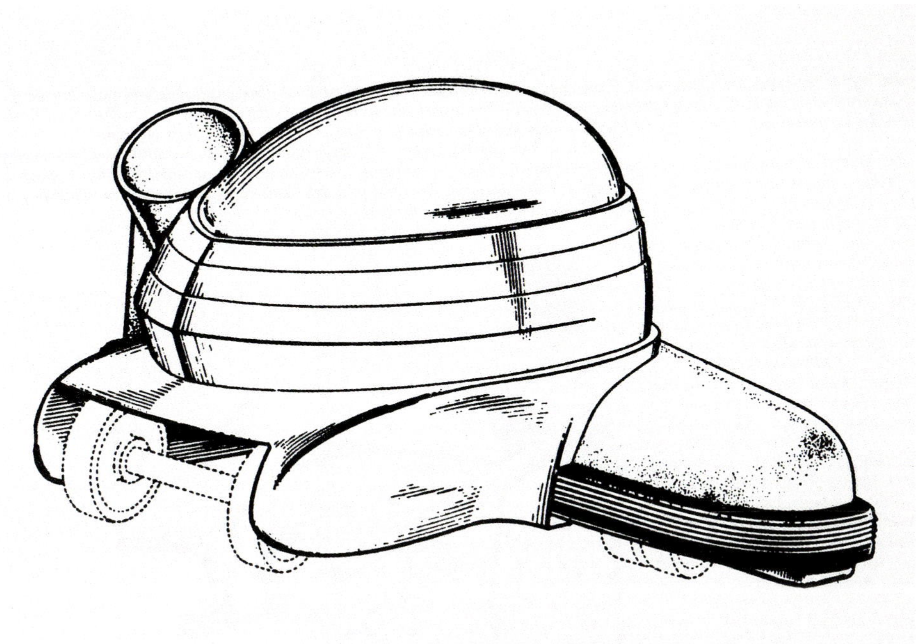

facing page

ADVERTISEMENT SHOWING A PHILIPS RADIO-SET EQUIPPED WITH A BAKELITE "SAUCER" LOUDSPEAKER, THE NETHERLANDS, 1927.

right, top

PATENTED MODEL FOR A HOOVER VACUUM CLEANER, UNITED STATES, 1930S–40S.

right, bottom

DOMESTIC BAKELITE AND CHROME-STEEL IRON, UNITED STATES, 1930S.

and that changed radically during the first decades of the twentieth. Shortly after World War II, phenolic resin was even adopted as a television receiver body. This material had many advantages: the temperature of approximately 480°F (250°C) at which it is worked is unlikely to damage the details of a steel mold. It could also be polished to a high degree, thus enhancing the finished product's formal qualities with its highly reflective sheen. Similar considerations furthered the prospects of a metal such as aluminum, whose melting point of 1220°F (660°C) leaves the shine of the impression of the steel mold intact. It is hardly a surprise, then, that both Bakelite and aluminum became particularly popular in the United States in the 1930s for products that embodied the new Streamline style. This stylistic trend had repercussions on domestic objects as forms originating in aerodynamic research were applied to motor vehicles as part of the effort to increase speeds by improving fuel economy. Thus vacuum cleaners and domestic irons dazzled the eye, looking like shiny racing-cars with rippling curves.

But these were not the only materials essential to pre-World War II style in the United States: pressed sheet-steel was used to create many iconic designs of the time. All curves, and awash with gloss cellulose paint and gleaming chrome, it gave life and soul to true-blue American glories such as those designed by Harley Earl for General Motors (Buick Y-Job, 1937) and Raymond Loewy for Studebaker (Studebaker Champion, 1938), as well as to many a make of carpet-sweeper, light-fitting, and electric toaster.

More than ever, steel was king. Its production was an affirmation of the might of the same industrialized nations that were soon to clash militarily. Between the wars, steel became all-pervasive in every sector of production. Few design innovators resisted the temptation to embrace it—folding and bending it, stamping and pressing it, or else employing it in the form of a semi-finished product that was to become wildly popular throughout the Modern movement: metal tubing. If Marcel Breuer was the first in 1925 to design a seating structure in curved steel tubing for his leather armchair B3, many other creatives soon embarked on similar ventures. Forsaking bentwood and instead adopting this semi-product much celebrated in avant-garde circles, the firm of Thonet was soon to become a kind of experimental laboratory. Models designed by pioneers such as Breuer, Mies van der Rohe, and the team of Le Corbusier/Perriand/Jeanneret were developed there and, launched on the market, appeared in the company catalog.

CELLULOID HANDBAG, FRANCE, 1930.

following pages
RAYMOND LOEWY IN HIS STUDIO WITH A PHOTOGRAPH OF THE S1 ENGINE FOR THE PENNSYLVANIA RAILROAD COMPANY (1937), TOGETHER WITH A MODEL CAR FEATURING IN A FURTHER STUDY IN AERODYNAMICS, UNITED STATES, LATE 1930S.

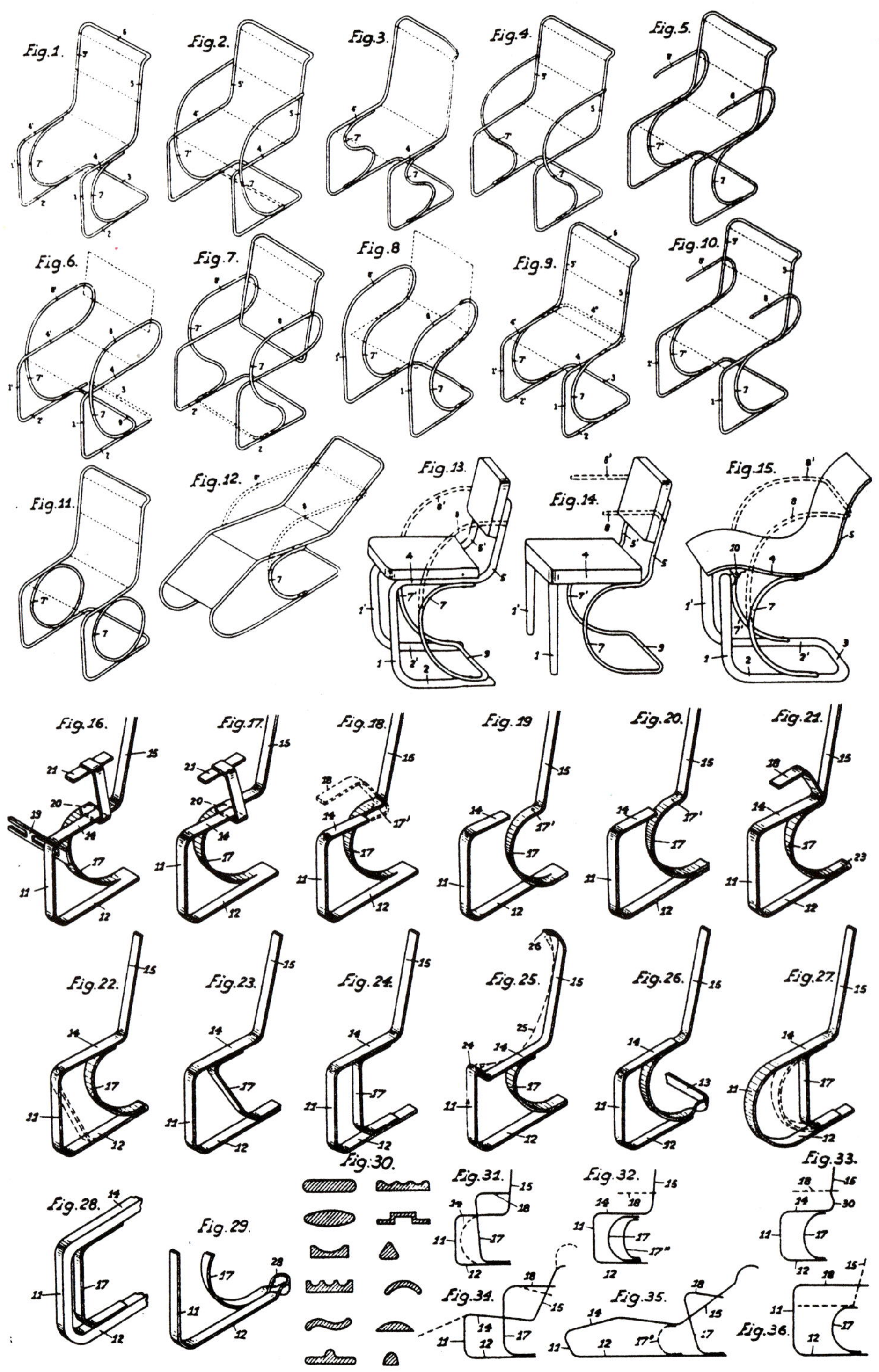
Fig.1.
Fig.2.
Fig.3.
Fig.4.
Fig.5.
Fig.6.
Fig.7.
Fig.8
Fig.9.
Fig.10.
Fig.11.
Fig.12.
Fig.13.
Fig.14.
Fig.15.
Fig.16.
Fig.17.
Fig.18.
Fig.19
Fig.20.
Fig.21.
Fig.22.
Fig.23.
Fig.24.
Fig.25.
Fig.26.
Fig.27.
Fig.28.
Fig.29.
Fig.30.
Fig.31.
Fig.32.
Fig.33.
Fig.34.
Fig.35.
Fig.36.

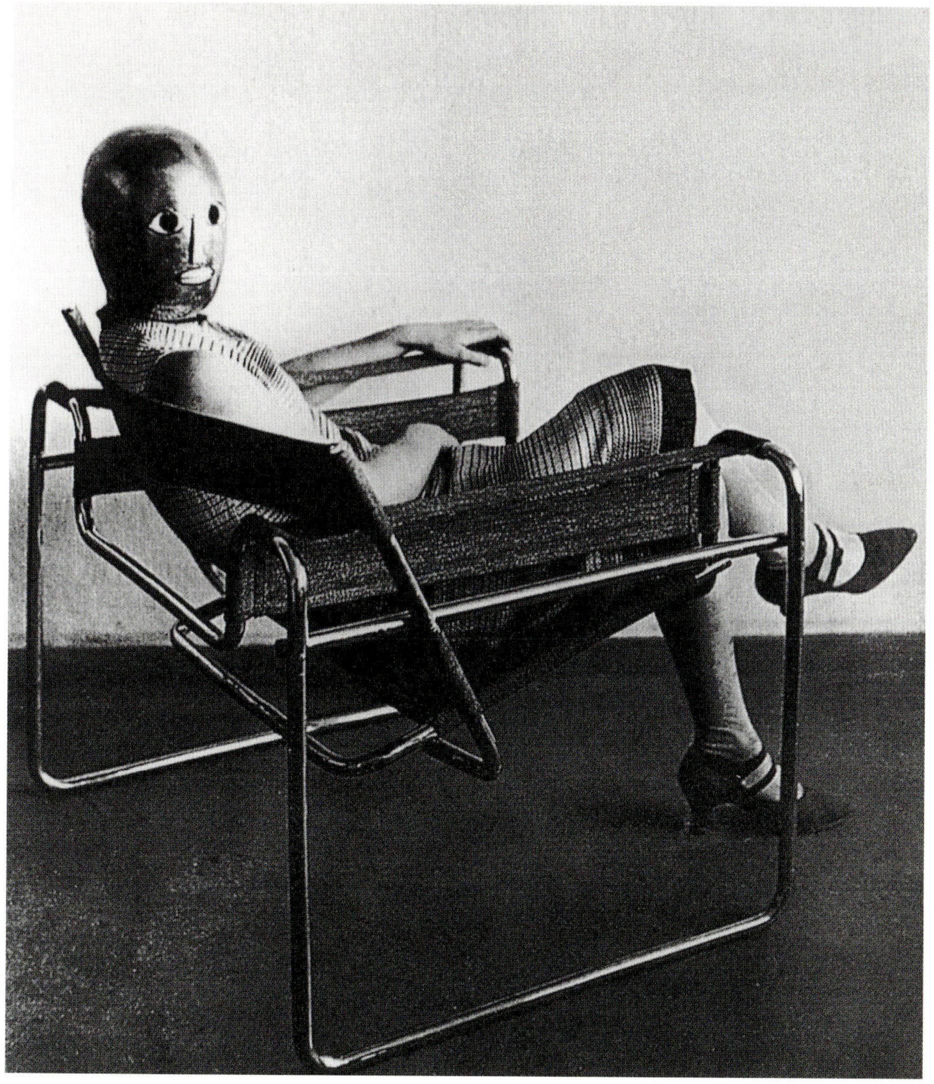

facing page

MARCEL BREUER, DRAWINGS OF PATENT NO. 170986 LODGED IN 1933 SHOWING, AS HE CLAIMED, HOW THE STRUCTURE FOR A BENT TUBULAR-STEEL SEAT COULD BE TRANSFORMED INTO A STRUCTURE IN STEEL SECTIONS WITH ALUMINUM SIDE PIECES.

WOMAN WEARING A MASK BY OSKAR SCHLEMMER SEATED IN MARCEL BREUER'S LEATHER EASY-CHAIR CLUB B3, THE DESSAU BAUHAUS, GERMANY, 1927.

following pages

CHARLOTTE PERRIAND IN THE TILTING LOUNGER BY LE CORBUSIER, JEANNERET, PERRIAND, ISSUED BY THONET, FRANCE, 1928.

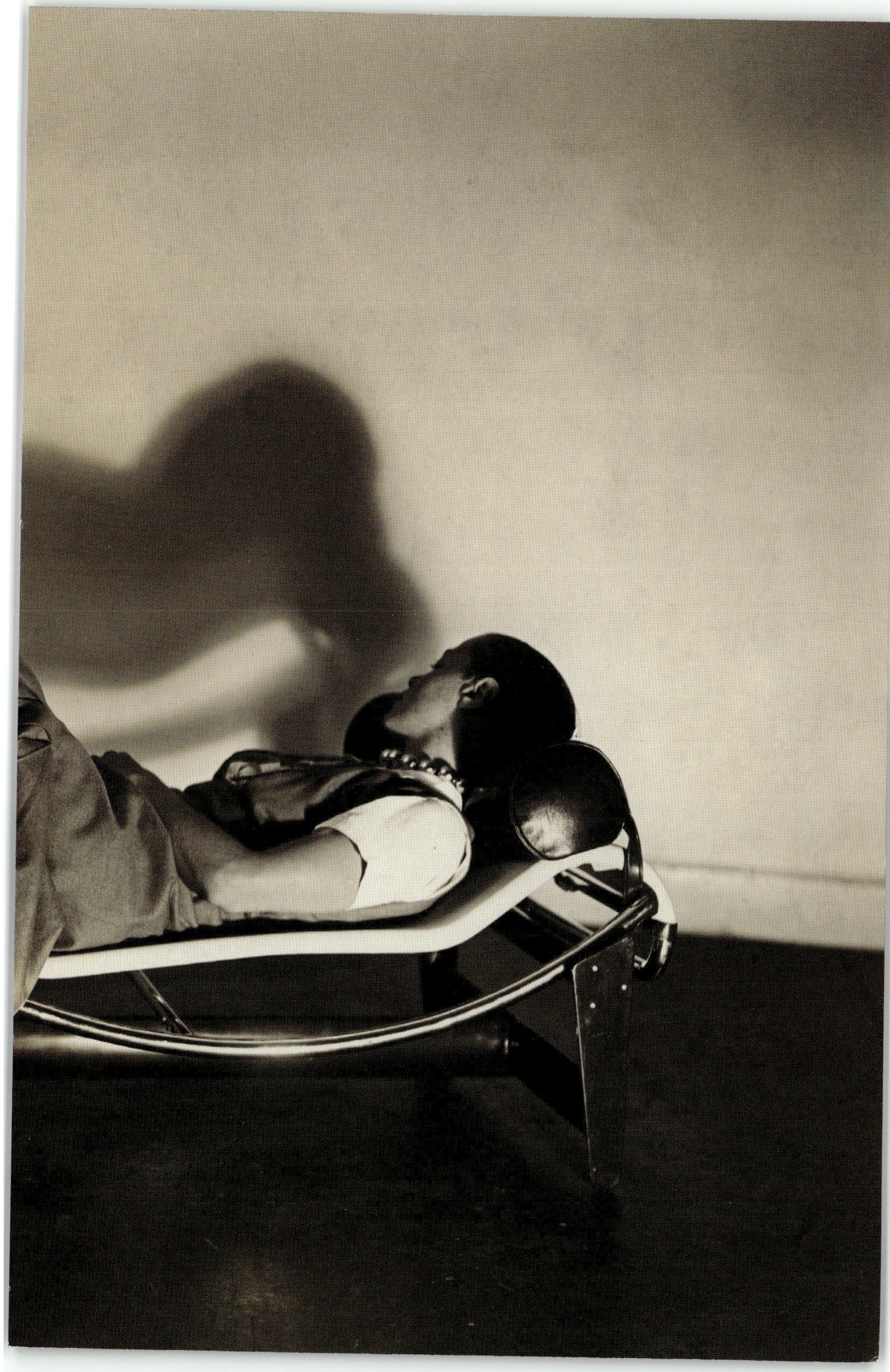

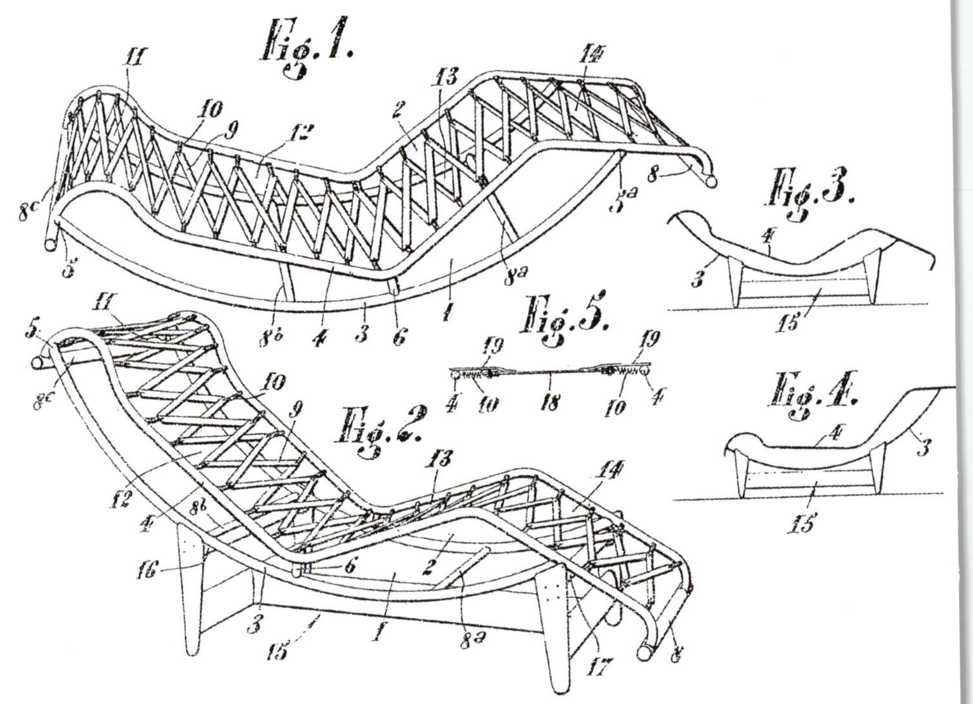
Fig.1.
Fig.2.
Fig.3.
Fig.4.
Fig.5.

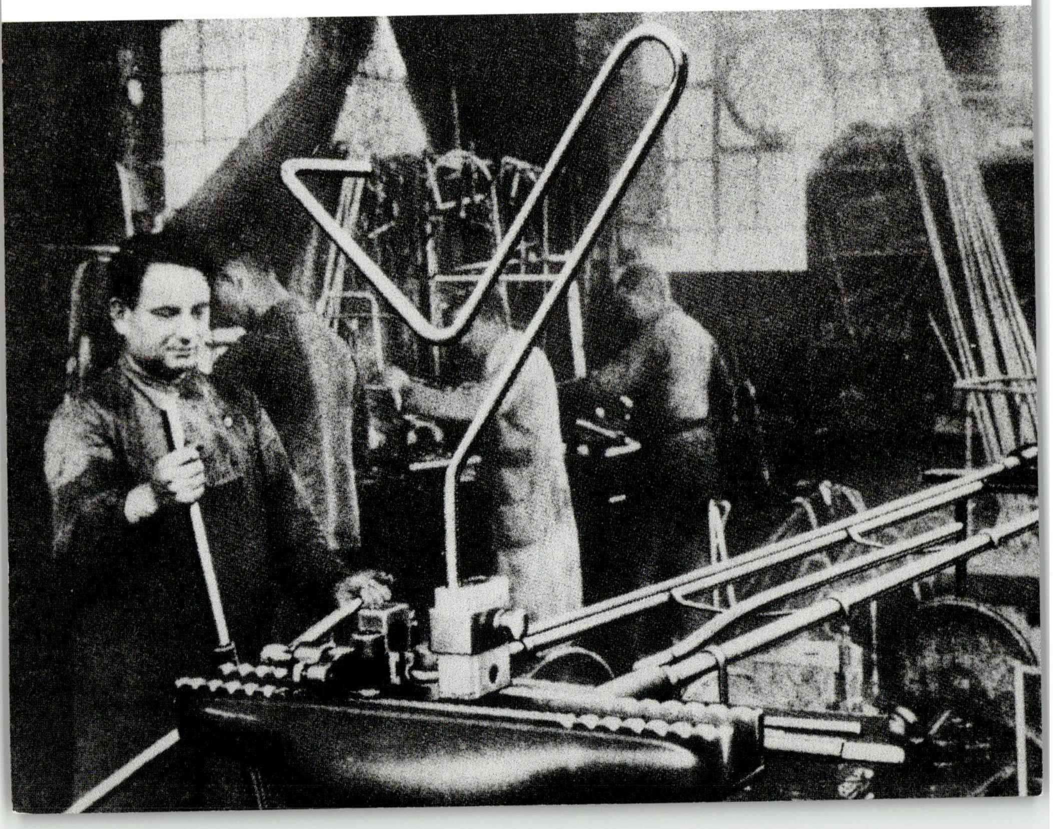

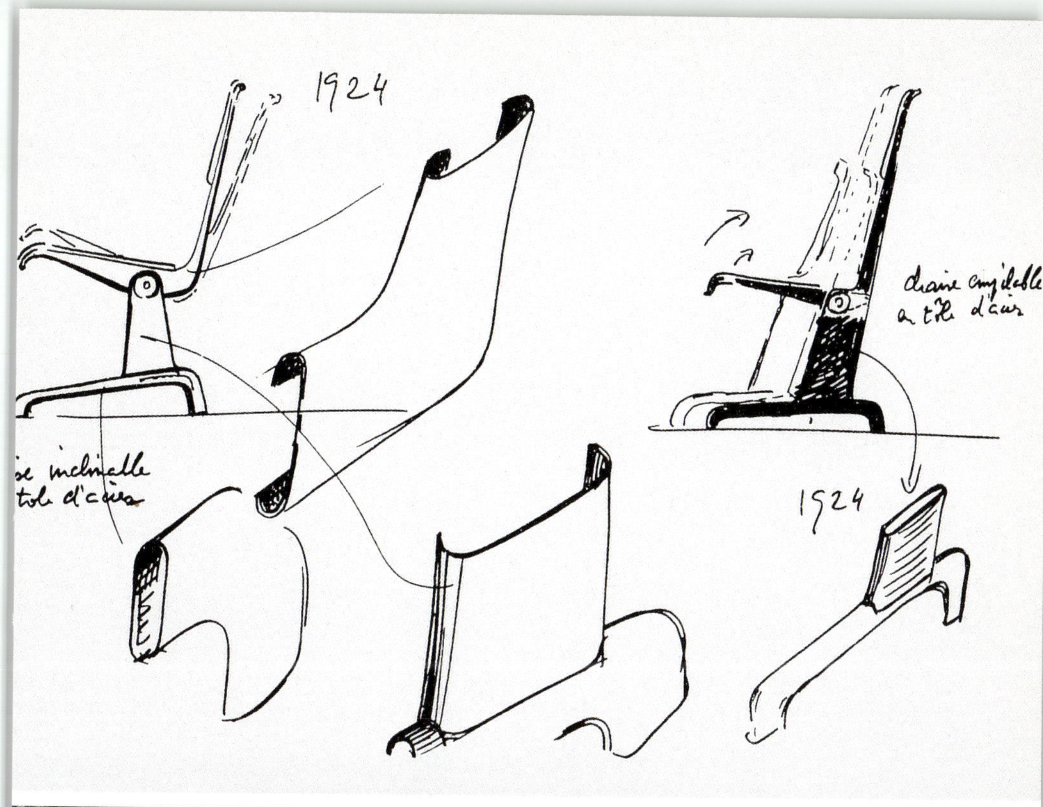
1924
1924

A3501
CONVERSATION
THICKNESS OF PLYWOOD VARIES
WITH STRUCTURAL DEMANDS

preceding pages

left, top

DRAWING OF A PATENT RELATING TO THE SLIDE IN THE TILTING LOUNGER BY LE CORBUSIER, JEANNERET, PERRIAND, LODGED APRIL 8 1929.

left, bottom

PRECISION BENDING OF STEEL TUBE AT THE CARL BECK & ALFRED SCHULZE CO. PLANT, GERMANY, 1930.

right, top

JEAN PROUVÉ, DRAWINGS FOR THE ARTICULATED CHAIR AND THE FOLDING CHAIR EXECUTED IN 1924 FOR THE MAGAZINE *INTÉRIEUR*. AS THE SEATS WERE PRODUCED AT THE END OF THE 1920S, THE DRAWINGS SHOW THE CONCEPT WAS ALREADY VIABLE IN 1924.

right, bottom

JEAN PROUVÉ, ARTICULATED CHAIR AS EXECUTED IN HIS WORKSHOPS AT NANCY, C. 1927–30.

facing page

CHARLES EAMES AND **EERO SAARINEN**, ONE OF THE CHAIRS IN THE WINNING PROJECT AT THE ORGANIC DESIGN IN HOME FURNISHINGS COMPETITION ORGANIZED BY THE MUSEUM OF MODERN ART, NEW YORK, 1940.

During the interwar years, wood still played a major role, though its more convincing results were in the guise of derivative products. On the design side, there was the remarkable use of glued-laminated timber and plywood bent and joined into a single unit by Alvar Aalto in the superb Paimio chair in 1931, the same period in which he devised a glued-laminate structure for a chaise longue featuring an impressive cantilever. Molded plywood also appears in seating shells by Charles Eames and Eero Saarinen that won awards at the competition "Organic Design in Home Furnishings" staged in 1940 by MoMA (the Museum of Modern Art) in New York. In the area of military research, the duo of Charles and Ray Eames developed the technique medically to manufacture stretchers and splints.

MILITARY ADVANCES IN PEACETIME

Military research is often the driving force behind advances in science and technology, and during World War II this was very much the case, not least with steel. Methods of continuous electric welding, devised to speed up the production of cargo ships and landing craft, had a knock-on effect on the assembly of elements for superstructures and building frames. Very quickly and in many sectors (transport machines, civil engineering materials, frames and housings), riveting was dropped in favor of mechanical welding.

It was perhaps in the field of chemical engineering that progress was most noticeable. If not every synthetic material was born from the war effort, all gained from it. Competition between the two giants of the sector, IG Farben in Germany and Du Pont de Nemours in the United States, was unrelenting and, as the advantage seesawed between the two firms, the drive to catch up become a further spur to research. Once peace returned, the public at large did not have to wait long to reap the fruits of this battle of the Titans. In 1946, in an exhibition in London entitled "Britain Can Make It," a section called "War to Peace" hailed the beneficial effects of military research on everyday life. Naturally enough, synthetic materials were very much to the fore, ushering in the Age of Plastic.

The majority of the synthetic materials that were being developed at this time were thermoplastics. These materials are better than thermosets, whose polymerization rate is relatively slow.

Injection molding, where there is only time for the accelerated cooling of the material introduced into the mold in the liquid state, allows for rapid throughput of finished objects. Some of these thermoplastics had already met with commercial success in the years before World War II. This was notably the case with polyamide 6-6, which triumphed in the United States under Du Pont de Nemours's trademark of Nylon from 1938 onwards in the form of stockings that proved both lighter and more sheer than those made in natural silk. A few years later, Nylon was used in parachutes.

Polymethylmethacrylate, developed in 1927 by Otto Röhm and marketed in 1934 in the form of sheets by the company of Röhm & Haas, and dubbed Plexiglas, was turned into airplane cockpits. At the turn of the 1940s, it was transformed into bedroom furniture for Helena Rubinstein who was bewitched by the transparency of the material and the curves into which it could be thermoformed. Other thermoplastics were known before the conflict broke out, but their applications were often for military purposes and, thus, secret. This was true of the polyamide IG Farben called Perlon, the German equivalent of Nylon, and also with polyethylene, which was developed by the British firm ICI to the point of industrial production on the eve of the war and whose license the company ceded to the United States. Flexible and completely neutral chemically, polyethylene proved more effective for processes such as sheathing underwater cables than gutta-percha. Its remarkable history in civilian garb is well documented, beginning with postwar advances in the highly flexible Poly-T, thought up in 1942 by Earl W. Tupper. In 1945, this onetime Du Pont de Nemours chemist had the idea of molding small airtight boxes whose lids would click into place on the container. Specially devised for refrigerating food, Tupperware, at first sold in the United States at "home-sale parties," is an incredible success story. In 1948, Great Britain saw the advent of polyethylene buckets and basins mass-produced in bright colors, which sold like hot cakes. "Flexible-hard" materials had entered the home.

In Germany in 1953, Karl Ziegler came up with a more rigid high-density polyethylene. This could be molded into furniture whose impermeability and shock-resistance made it ideally suited to gardens. This material also seemed tailor-made for children, and in 1964 Marco Zanuso and Richard Sapper designed a stackable chair for nursery schools brought out by Kartell. Sufficiently resistant to be load-bearing, well suited to bright colors, polyethylene was also blown into toys, in particular those equipped

BRITAIN CAN MAKE IT EXHIBITION,
WAR TO PEACE SECTION, LONDON,
1946.

following pages
WORLD WAR II DECOY TANK
IN RUBBERIZED FABRIC,
GREAT BRITAIN.

760

THE STYRENE SONG

Oh, time, suspend your bowl, oh plastic substance
Whence are you? Who are you? And what accounts
For your rare qualities? Of what might you consist?
From what have you come? The object, as it now exists,
Has remote ancestors! Let us in reverse unfold
Its exemplary history. To begin with, the mold.
Including the nest, mysterious entity,
It engenders the bowl or whatever else may be.
But the mold is itself enclosed in a press
That injects the paste and forms the piece,
And thus most advantageously extrudes
The finished object, ready to be used.
The mold is expensive; this could be a prohibition.
But then it can be rented out, even to the competition.
Vacuum forming is another way of
Obtaining objects: simple aspiration pays off.
The warmed material, carefully gauged,
Is extruded in sheets at the previous stage.
To go into the injector a piston was needed, so
Too the isomantle—or the mantleiso—
Into which was fed ... What? Polystyrene,
Hardy and boisterous, rushing, not serene.
Vibrating on the sieve, the granulated swarm
Bustled happily in colors so warm.
Before being a granule it was an extruded string,
Strings of all tints, shades, tones, coloring.
These strings had been, following a procedure,
A sausage to which a plasticating screw applied pressure.
And what occasioned the agglutination?
Pearls colored to suit every imagination.
And how was it colored? Why, the pigment mixed
With polystyrene became homogeneously fixed.
But before that the product to dry was tumbled
And so, rotatingly, the product stumbled.
It was scarcely born, our polystyrene.
A polymer produced by the simplest styrene.
Polymerization: this word, as we all know,
Designates obtaining a complex product, and so
Of higher molecular weight. And in a reactor,
An elementary machine, work of an engineer,

The molecules into pearls, clinging, formed
And linked to each other. Yes—but before?
The styrene was only a liquid, colorless,
Somewhat explosive and not odorless.
Do take a good look: it's the only occasion
You'll have to see what is in question.
Styrene is produced in great quantities
From superheated ethyl-benzenes.
Formerly styrene was extracted from benzoin,
Product of the styrax, in Indonesia grown.
And so from pipe to pipe we are going back,
Through the desert of canalizations' track,
Toward the raw materials, toward matter abstract
Circulating without end, effective and secret.
It is washed and distilled and then redistilled
And these are not mere stylistic exercises skilled:
Ethylbenzene can—and even should blow up
If its temperature rises high enough up.
As for ethylbenzene, it comes, I have seen,
From the combination of liquide benzene
With ethylene, a simple vapor.
Ethylene and benzene have as generator
Either coal, or oil, or oil or coal.
To make them both, they both can have a role.
We could set off down this new avenue
To try to get of both of their origins a view.
Does oil come from masses and masses of fish?
It's not really known, and coal's origin's a wish.
Does oil come from plankton in labor?
A much debated question ... growing ever grayer...
And oil and coal went up in smoke
When along came the chemist who, with luck's stroke,
Turned these clouds solid and made of them
Innumerable objects utilitarian.
Into new materials these obscure residues
Are thus transformed. And unknown brews
Are still awaiting chemical mutation
So they can merit commercial exploitation.

RAYMOND QUENEAU, 1957-1958

with casters on which children could sit. Additionally, PET tubing could be blown to form bottles.

In 1954, Giulio Natta in Italy developed another moldable material, polypropylene, which was endowed with a measure of flexibility and could be dyed to a fine translucency. Remarkably tough, it was to become a serious competitor to polyethylene. In 1963, Robin Day was among the first to use it for what became a popular chair, the Polyprop. Today polypropylene is one of the synthetic materials most appreciated by designers, particularly Philippe Starck, who has employed it in many chairs, including Bubu I and Lord Yo (1992), Cheap Chic (1997), and Toy (1999). Polychloride Vinyl (PVC) also saw the light of day in the prewar period and was first produced in the early 1930s by Union Carbide in the United States and by IG Farben in Germany. It replaced rubber as an insulator for electric cables, its main use during the war, and was soon employed as a coating for weatherproofing protective clothing fabrics.

After the war, flexible and extruded PVC continued to be deployed as a material for whipping electric cables and wires but was also made into pipes and washers of all kinds. But it was in the form of calendered products that it was to become one of the most prized materials among designers in the 1960s. In the form of thin, non-flammable sheets—colored or plain, transparent or opaque, and easy to high-frequency weld—it was the inspiration behind many inflatable designs. In avant-garde circles, many architects, working individually or in architects' practices, dreamt up and even sometimes realized inflatable structures in PVC. Toward the end of the 1960s, designs in Italy (the Blow chair by Scolari, De Pas, D' Urbino, and Lomazzi) and France (pieces by Bernard Quentin, the Aerolande group, Quasar) focused on furniture. In 1968, Quasar thought up a furnished cell whose components (flooring, partition, ceiling, couch, armchair, and light-fittings) were inflatable. In the same years, couturier André Courrège introduced PVC sheeting into fashion design.

A WOMAN SITTING ON THE CURB AND PUTTING ON THE PAIR OF NYLON STOCKINGS SHE HAS JUST PURCHASED, WHILE THE LONG LINE IN FRONT OF THE STORE LOOKS ON IN AMUSEMENT, UNITED STATES, 1940S.

Another semi-finished product obtained by coating textile in PVC, Skaï (or leatherette), a substitute for hide or leather, was much taken up for fine leather goods and seat cushion covers. Gatti, Paolini, and Teodoro were to employ it to wrap their astonishing Sacco chair (1968).

Polystyrene was another thermoplastic born in the prewar period, but was little known at the time. Developed in 1933 by

TUPPERWARE PARTY IN FLORIDA,
END OF THE 1940S.

LANDISLAS MEDGYES, BEDROOM FURNITURE IN LUCITE (ACRYLIC RESIN) FOR HELENA RUBINSTEIN, UNITED STATES, 1940.

following pages

left

ROBIN DAY, POLYPROP CHAIR. THE SHELL INCORPORATING THE BACK AND SEAT WAS THE FIRST TO BE MOLDED OUT OF POLYPROPYLENE. BROUGHT OUT BY S. HILLE & CO, GREAT BRITAIN, 1962.

right

PHILIPPE STARCK, LORD YO ARMCHAIR, THE SHELL FOR THE BACK AND CHAIR MOLDED OUT OF POLYPROPYLENE, PRODUCED BY DRIADE, ITALY, 1994.

a German professor, Herr Wulff, it was initially molded into small objects of a technical nature. During the war, progress was advanced significantly by the massive production of styrene necessary for the manufacture of synthetic rubber. With peacetime, polystyrene quickly came to the fore in many industrial sectors, particularly those concerned with the mass manufacture of homeware. Highly thermoplastic, it is easy to mold and was used to produce objects of every kind: utensils and articles for the table, as well as housings and cases for domestic appliances, radio sets, etc. It was competing with acetate and initially was a fragile material and of inferior quality. Its poor reputation resulted in the gradual creation of a polystyrene of superior quality—high-impact polystyrene (HIPS)—developed during the war by IG Farben. But its problems really came to an end with ABS (Acrylonitrile-Butadiene-Styrene) in which a synthetic rubber, polybutadiene, is grafted onto the hard styrene acrylonitrile resin, bestowing a certain flexibility and remarkable impact resistance. ABS molds easily and, dyed in the mass, offers a very wide range of colors. By the 1960s, it had become omnipresent, used for everything from housings for telephones, to bodies for electrical home appliances, typewriters, radio sets, record players, seats, or into components for small storage units. It could even be thermoformed and turned into a fascia element for a piece of furniture, a store sign, or as in the case of the Citroën Méhari designed by Jean Louis Barrault in 1968, an automobile body.

The Functionalist trend of the 1960s seemed especially attracted to black and white, sometimes set off by orange accents. At that time, a host of objects molded in ABS became available in these colors, as many items produced by Braun bear witness. Italian designers reacted vigorously against these sober tendencies, with many standard products parading far less reticent liveries. Whether the housing for the little transistor radio TS502, designed by Marco Zanuso and Richard Sapper for Brionvega in 1964, came in yellow, red, or green, it always did so with brio. The Valentine typewriter designed by Ettore Sottsass for Olivetti in 1969 flaunted an eye-popping red. By the end of the 1960s, this taste for bright colors had become widespread in France, where ABS tended to make its mark with somewhat raw hues, as exemplified by the Méhari.

preceding pages
AEROLANDE GROUP (JEAN AUBERT, JEAN-PAUL JUNGMANN, ANTOINE STINCO), INFLATABLE FURNITURE IN FLEXIBLE VINYL POLYCHLORIDE, FRANCE, 1967.

facing page
ANDRÉ COURRÈGES, RAINCOAT AND BOOTS IN SOFT VINYL POLYCHLORIDE, FRANCE, 1969.

following pages
THREE-DIMENSIONAL TEXTILE (POLYESTER, POLYAMIDE, POLYPROPYLENE) COMPOSED OF LENGTHS OF FABRIC CONNECTED BY WIRE. THE DISTANCE BETWEEN THE STRIPS CAN RANGE FROM A FRACTION OF AN INCH TO 18 IN. (450 MM). THESE TEXTILES ARE EMPLOYED AT PRESENT IN INFLATABLE STRUCTURES AND FOR HEAT-RESISTANT ELEMENTS. THEY CAN ALSO BE MADE INTO INFLATABLE FURNITURE. MANUFACTURED BY TISSAVEL, FRANCE.

MARCO ZANUSO AND

RICHARD SAPPER, RADIO-SET,

TS 502 BRIONVEGA (CASE IN ABS),

ITALY, 1964.

GROUNDBREAKING MATERIALS

Among all the thermosets that appeared before the war and that, just like thermoplastics, had advanced in leaps and bounds thanks to military research, two made a particular impact: polyurethane and polyester. At the end of the 1930s, Otto Bayer, who had been laboratory head at IG Farben from 1934, developed a practical reticulated polyurethane that was made workable in the form of a rigid foam in 1941. Its lightness saw it used particularly as a replacement for balsa wood in aviation and as heat insulation in German submarines. In 1950, Bayer also obtained the first elastomers and flexible polyurethane foams. Poorly resistant to damp over time, they were significantly improved by Du Pont de Nemours in 1953 when one of its components, polyester, was replaced by polyether. In addition to possessing superior aging characteristics, these new foams were also more elastic and thus more similar to the latex foam with which they began to compete. Readily moldable into trimmings for car seat cushions, for example, they mainly retailed in the form of large-sized blocks. They were cut into slices of various thickness and made into elements for upholstery, mattresses, etc. To a significant degree, it was the advent of the large-scale production of flexible polyurethane foam that spawned the "floor-hugging" lifestyles of the 1960s. Covered with brightly colored fabric, laid straight on the floor, and easily moved about, foam blocks soon ousted beds and chairs. Foam also had a part in more orthodox seating, however, as not everyone was prepared to follow this trend, and even its adherents found armchairs, settees, and chaise longues a comfortable option from time to time. There, too, foam made its mark and affected the development of forms that soon became a style. Designed by creative figures such as Pierre Paulin, Verner Panton, and Olivier Mourgue, seating was constructed from tubular substructures supplied with thin rubber straps covered by a layer of foam—latex or polyurethane—with the whole unit covered in jersey. The foam was rounded off at the corners, where it was held in place by the cloth, giving the piece those supple, smooth forms so characteristic of the period. Occasionally, a block of foam was trimmed with a hot wire into the shape of an entire chair, as with the Malitte, designed by the painter Roberto Matta for Gavina in 1966. This was also the case with the modular couch Superonda, conceived by the Archizoom Associati group and marketed by Poltronova, also in 1966. By 1991, Ron Arad's Soft Big Easy,

following pages
FACTORY PRODUCING
POLYURETHANE FOAM SHEETING

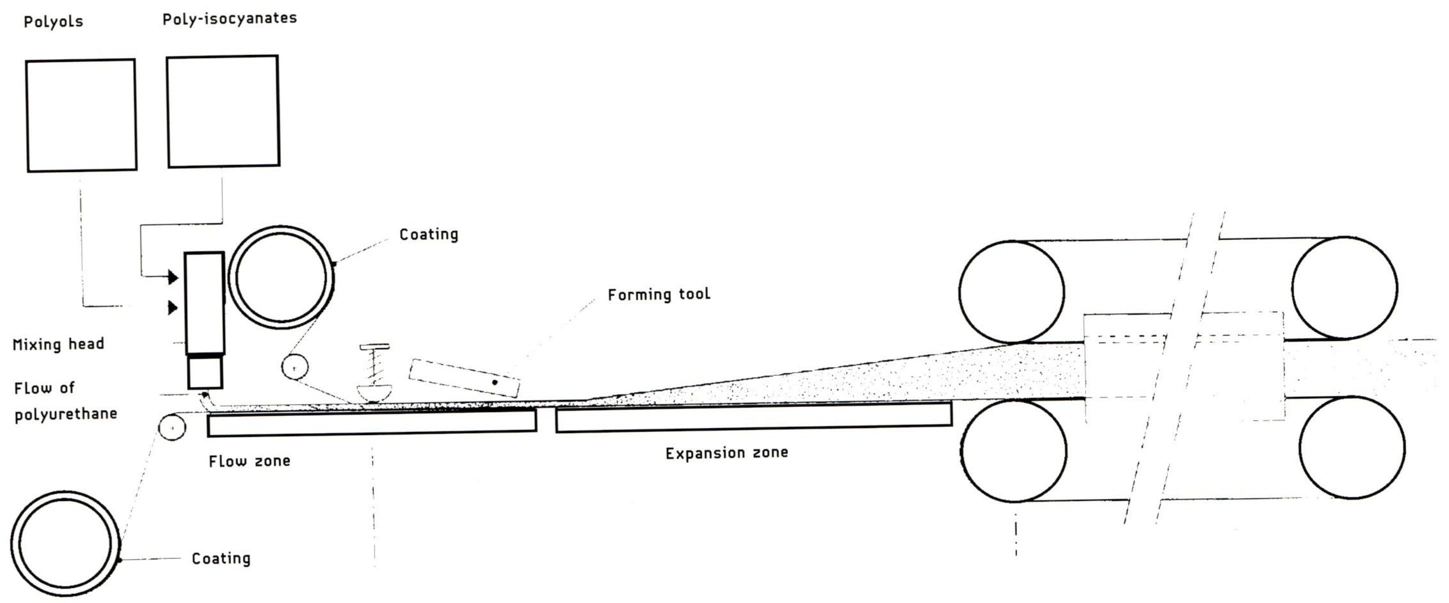
Polyols
Poly-isocyanates
Coating
Forming tool
Mixing head
Flow of
polyurethane
Flow zone
Expansion zone
Coating

issued by Moroso, was carved out of a foam block by a hot wire that was computer controlled. Polyurethane foam, though, is a lot of air and not much substance, a fact that in 1969 exercised the imagination of Gaetano Pesce in his UP series of seats. To lower the volume when the product was transported from the production site to the distribution point as well as in storage, the seats, molded out of a single piece of polyurethane foam, were placed in a vacuum-packing machine. They emerged in the form of flat-packs protected from the atmosphere by a PVC film that just has to be pierced for the air to rush in, inflating the foam and returning the seat to its original volume.

In 1936, Ellis Foster developed unsaturated polyesters that were conducive to low- or even zero-pressure molding in the open air, hardening at ambient temperature. Though without exceptional mechanical qualities, these polyesters were perfect insulators and were initially used largely as substrates for electric circuitry and in protective varnishes. It was during the war that they acquired, mechanically speaking, their letters of nobility, though not because the resin itself was improved, but because in 1942 United States Rubber had the excellent idea of reinforcing it with fiberglass (produced since 1937 by Owens Corning Fiberglass), a material displaying remarkable resistance to traction. Reinforced with fiberglass, polyester could be manufactured into sturdy structural elements and, in consequence, could vie with aluminum alloys and even steels. Initially, it was primarily employed in the manufacture of airplane radomes.

After the war, applications for fiberglass-reinforced polyester became more varied. The relative facility with which it could be handworked into small- or large-size parts made it suitable for constructing prototypes to tight deadlines. In 1957 at the Salon des Arts Ménagers in Paris, Charbonnages de France presented a prototype for a house designed by Ionel Schein in plastic, where considerable space was given over to fiberglass-reinforced polyester (including for the furniture). In the industrial field, flat and corrugated plates were form-molded or impressed continuously on the table for the construction industry, as well as in segments for pleasure boats worked by simultaneous projection (by spray gun) of resin and fiberglass, car bodywork elements (the roof-head on the DS19 Citroen in 1955, for example), and furniture.

DIAGRAM SHOWING THE MANUFACTURE OF POLYURETHANE FOAM.

The seating shells designed after the war by Eero Saarinen for Knoll and by Charles and Ray Eames for Herman Miller are especially intriguing. The famous molded-plywood seat shells of Eames and Saarinen were ergonomically adequate, but the techniques implemented were complex, slow, and costly.

Polyester strengthened with glass mat offered both designers a solution applicable to their respective seat shells. Eero Saarinen's Womb chair, designed for Knoll in 1946, features a shell covered in fabric. The Eameses do not take this approach for their Plastic Shell group, created for MoMA's "International Competition for Low-Cost Furniture Design" of 1948. Herman Miller, the future manufacturer of the series, sponsored the Eameses, and the shells were realized by Zenith Plastics, during the war a manufacturer of aeronautical radomes in polyester and fiberglass. The shells were delivered to the points of sale in their raw state, translucent with the glass mat remaining visible. Hard to dye in the mass, the polyester was lightly pigmented (in ocher, yellow, blue, or green). There was one major drawback: the risk over time that the fiberglass might leak. By 1955, bright colors and measures to prevent the fibers from leaching were devised at Knoll for Saarinen's Tulip chairs. Pre-coating the mold with a resin known as Gel Coat blocked the fibers and, by covering the product, also acted as a lacquer allowing for a wide spectrum of colors. If this was not exploited in the Tulip series itself, which stuck to white, strikingly colored reinforced polyesters soon became the height of fashion. Key examples include the No. BA1171 chair drawn in 1964 by Helmut Bätzner for Bofinger (the first seat in synthetic material molded in a single part conceived for mass production), Eero Aarnio's Pastille (1968), or the Molar chair by Wendell Castle (1969). This chromatic vitality combined with unbridled formal freedom revealed the influence of Pop Art and American West Coast Psychedelia.

The major role of fiberglass-strengthened polyester in the creation of new products after the war is, in fact, only part of a more general effect of material types whose economic importance continues to grow even now: composite materials.

VERNER PANTON, UPHOLSTERED SEATING SYSTEM, CURVED TUBULAR-STEEL CHASSIS, FOAM STUFFING COVERED IN JERSEY, DENMARK, 1963.

IONEL SCHEIN, ALL-PLASTIC HOUSE, MADE FOR CHARBONNAGES DE FRANCE AND UNVEILED AT THE SALON DES ARTS MÉNAGERS, PARIS, 1956.

UNITED THEY STAND: THE STRENGTH OF COMPOSITE MATERIALS

Today, the term "composites" refers primarily to materials that combine a reinforcing element, in general fibrous, with a matrix with which it is coated or enwrapped. However, this is a somewhat restrictive definition, and a more satisfactory, general one was proposed by B. Delapalme, at the time general manager of the Elf Rhône Alpes company, at the first international conference on composite materials held in Lyon in September 1971 by the Association Nationale de la Recherche Technique[1]: "the assembly, on whatever scale, of different bodies, with the aim of combining, with a view to improving quality, the various characteristics of the component materials."

According to this broader definition, depending on the project concerned, reinforced plastics (thermosets and thermoplastics), sandwich materials, laminates (stratified materials on kraft), plywood, fiber- and particleboards, concrete (reinforced or not), and even metal alloys can be considered composite materials in the full sense of the term.

Although artificial in origin, all these materials have their natural models. Composites with reinforcing fibers bound by resins recall the fibers of wood, its ligneous parenchyma and cells impregnated with gums and resins. The crisscross, folded structure of a mammal's stomach suggests plywood. The skin of an orange, with its compact external layer that conceals a flexible interior, or a dolphin, with its tightly interlocked skin covering a thick and spongy dermis, resembles certain flexible sandwich composites. Non-reinforced concrete, an aggregate of gravel and sand with a cement binder, constitutes a kind of stone that resembles granite (a magmatic rock formed by alkaline feldspar, quartz, and plagioclase). Consolidated into a tubular structure, it resembles a grass stalk. Such similarities lead to the field of bionics, a word invented in 1958 by Major Jack E. Steele of the aerospace medical division of the US Air Force. In his words, "Bionics is the science of systems which have some function copied from nature, or which represent characteristics of natural systems or their analogues."

following pages

CROSS SECTION OF A PIECE OF POPLAR EXAMINED UNDER A SCANNING ELECTRON MICROSCOPE (ENLARGEMENT 200X). THE VESSELS AND FIBERS OF THE WOOD ARE CLEARLY VISIBLE. AS THERE IS NO DISTINCTION BETWEEN THE SPRING AND SUMMER GROWTH, WOOD IS TERMED A "HOMOGENEOUS" SUBSTANCE.

1 I was a participant and "Composites and Design" was already the theme of my paper.

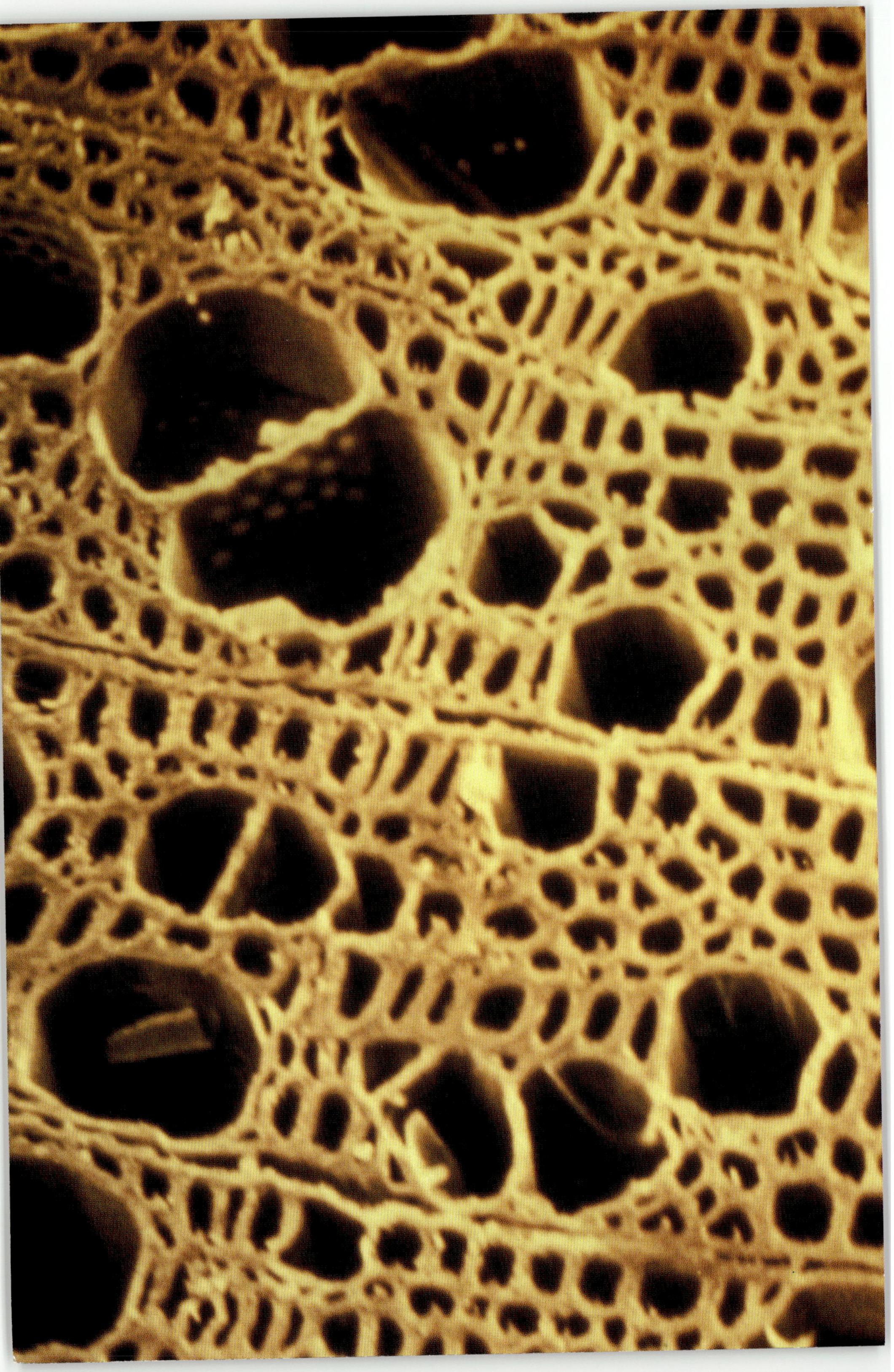

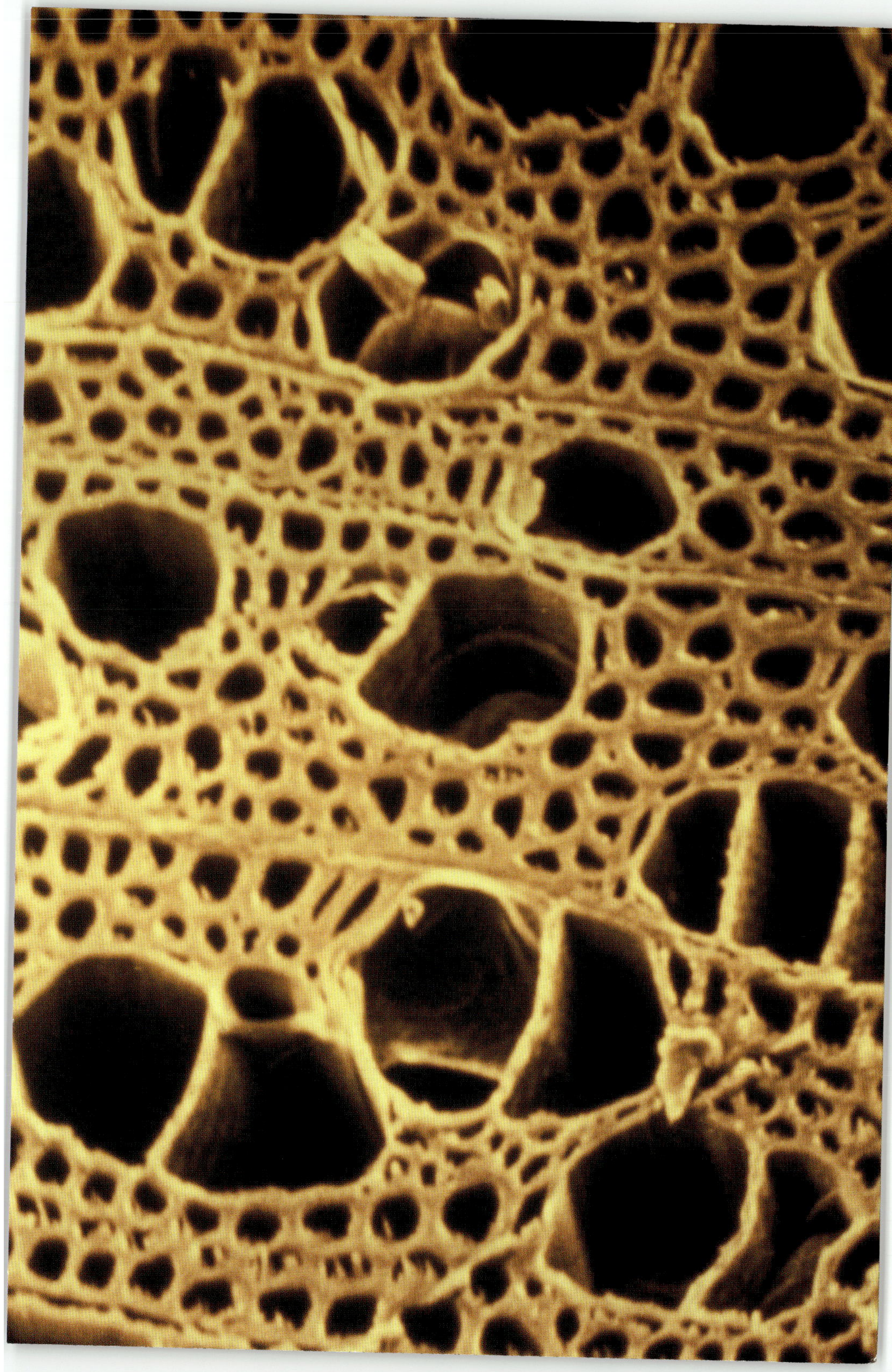

It seems that, knowingly or implicitly, the composites employed today are based on natural structures, in particular on those developed by life on Earth. Nature is thrifty, and in her constructions she places just the right amount of material exactly where it is required. For a long time now, the engineer's task has been to make optimal use of a material in a finished and potentially complex product, as seen, for example in electricity pylons and giant radiotelescopes.

In architecture of the postwar era, spatial structures (three-dimensional structures, stretched or inflated, geodetic lattices, domes, etc.) designed to make savings in material have became an essential facet of research. Today, the computer supplies rapid solutions to problems of material resistance that, in former times, were so arduous (if not impossible) to solve by the calculating methods available that architects had to resort to experimentation or analogy.

Research on these macrostructures is similar to that relating to composites. In the area of bionics, microscopic exploration of natural structures has opened doors to a magical and often orderly world. Think of how rigorously the material in a bone tissue, or in the skeleton of a sponge, radiolarian, or foraminifera, or in a fly's eye or pollen grain is organized when revealed under an electron-scanning microscope.

Today, nanotechnology proposes interventions on the atomic scale. The scanning-tunneling microscope developed in 1980 by the German Gerd Binning and the Swiss Heinrich Rohrer magnifies one hundred million times, thereby allowing observation of the atomic surface of a material and leading to the possibility in the near future of being able to act directly on the arrangement of atoms. Already in the field of composites, reinforcing structures capable of considerably altering the properties of certain polymers can be incorporated on the nanometric scale. This is a field of cutting-edge technology in which the designer, just like the engineer, is in a win-win situation. Experience proves that, when a new material arrives, designers are rarely the last to find applications for it, pushing its performance to the limits and prepared to reach for others once these are reached. Ample proof of this is afforded by the molded plywood in the prizewinning project of Eames and Saarinen in the 1940 MoMA competition, and after the war by the use by the same designers of reinforced polyester to facilitate the manufacture of seating shells. For a design, objectives should also focus unrelentingly on optimizing performance.

STALKS OF HORSETAIL,
A PLANT WITH HOLLOW STALKS
THAT GROWS IN DAMP FIELDS,
ALSO CALLED RAT'S TAIL.

RICHARD BUCKMINSTER FULLER SHOWING HIS MODELS FOR THREE-DIMENSIONAL STRUCTURES INCLUDING THAT OF A GEODESIC DOME SIMILAR TO THE 250-FT. (76-M) DIAMETER DOME IN THE UNITED STATES PAVILION AT THE UNIVERSAL EXHIBITION, MONTREAL, IN 1967.

ROBERT RICOLAIS WITH
THREE-DIMENSIONAL LIGHT
STRUCTURES IN HIS RESEARCH
LABORATORY AT THE UNIVERSITY
OF PENNSYLVANIA, 1960S.

Thus the quest for the lightest possible chair has been characterized by a number of landmark projects. The first was Gio Ponti's Superleggera, created between 1955 and 1957. Realized in solid ash whose sections were pared down to a minimum, it weighed scarcely more than 4 lb. (1.75 kg). In 1987, Alberto Meda and the technicians at Alias pushed a carbon fiber and a kevlar honeycomb core coated with epoxy resin to its limits to produce the Light-Light seat that weighs just over 2 lb. (1 kg). Though to date a record, its production costs were prohibitive. Patently not intended for general sale, this seat has become nevertheless a benchmark. With its 5 lb. (2.3 kg), the Laleggera chair designed in 1993 by Riccardo Blumer and made in 1996 by Alias cannot outdo Ponti's chair in terms of lightness, but its hollow solid maple structure with maple, wild cherry, or veneer injected with rigid polyurethane foam is the result of a novel manufacturing process. In 2000, François Azambourg came up with another impressive response in the form of the Pack chair, sold in a 3-US quart (3-liter) volume pack containing a rolled-up fabric sheath and polyurethane as filling. Once preset for foaming, by simply turning a button, the polyurethane expands into the envelope, where it hardens in five minutes.

At just under 4 lb., (1.7 kg), the Pack chair weighed as little as the Superleggera. But Blumer had not quite finished and turned to a natural material—balsa, a wood made up of large cells with extremely fine walls providing an incredibly low density of 0.15. Cut into elements for assembly, the unit was covered with resin and fiberglass to make, in 2004, a sturdy chair weighing 2 4/5 lb. (1.3 kg). The example of the balsa "limitEDition" chair is all the more interesting as it took no less than ten prototypes to achieve the final solution. The ten handmade prototypes prove that even today not everything in design can be solved by the magic wand of computing. But for how long?

ELECTRON MICROSCOPE
PHOTOGRAPHS OF A RADIOLARIAN,
A MARINE PROTOZOON WHOSE
FREQUENTLY SPHERICAL
SILICEOUS SKELETON POSSESSES
A DIAMETER OF THE ORDER
OF A MILLIMETER.

following pages

left

GIO PONTI, SUPER LEGGERA
CHAIR, WITH A SOLID ASH
STRUCTURE, MANUFACTURED
BY CASSINA, ITALY, 1957.

right

ALBERTO MEDA, LIGHT-LIGHT
CHAIR, IN CARBON FIBER, RESIN
POLYESTER, AND KEVLAR
HONEYCOMB, MANUFACTURED
BY ALIAS, ITALY, 1987.

CHAISE
CHAIR
P A C K
mode d'emploi
Directions

THE VIRTUAL IN A MATERIAL WORLD

For many designers, the language of computing has become indispensable, a fact already accepted in the 1980s by certain American agencies. Not least among them was Frogdesign, founded in 1969 in Germany by Hartmut Esslinger, who settled in the United States in 1982 to collaborate with Apple Computer (the Apple IIc microcomputer appeared in 1985) to exploit CAD (computer-aided design) and synthesized imagery. During the 1980s, this agency was joined by fledgling groups such as Design Logic and Technology Design, whose fluidity of line in some products betrays the impact of formal research on computer screen. In France in 1987, Marc Berthier produced pioneering designs on computer for an office chair. Since then, others have seized computing tools, sometimes to devise original implementations for a material. Martin Szekely's Armoire, issued by Kreo in 1998, was entirely conceived on and manufactured by computer.

Szekely used Alucobond, a sandwich made up of a polyethylene core with an aluminum faceplate on both sides. The sheet was cut and then incised—on one aluminum coating to allow for folding and then on both to create the hinge fitment—by a digitally controlled milling-machine.

In a few decades, both CAD and ICAM have made incredible strides.

Today, every computer on the planet can be interconnected via Internet and organized into networks. Digitized, any information (text, sound, image) can be loaded onto the network and, thanks to the increase in the number of geostationary satellites, received in real time whatever the distance from the point of emission. Better still, one can even, thanks to ICAM, imagine the product under study not simply circulating as a 3D image on a computer screen but, at certain stages of development, being received in the form of volume models that are built up layer by layer at a rapid prototyping unit.

The perspectives made available by technology in the future will be immense. In this context, and following early experiments by a trailblazer like Ron Arad in 1999, it is scarcely surprising to see other designers becoming interested in the potential of such processes. If industry, for the time being, employs them solely for making prototype parts that, to be

preceding pages

left

RICARDO BLUMER, LALEGGERA CHAIR, SOLID MAPLE, WITH MAPLE, WILD-CHERRY, OR ASH VENEER AND RIGID POLYURETHANE FOAM, PRODUCED BY ALIAS, ITALY, 1996.

right

FRANÇOIS AZAMBOURG, PACK CHAIR, FABRIC AND RIGID POLYURETHANE FOAM, FRANCE, 2000.

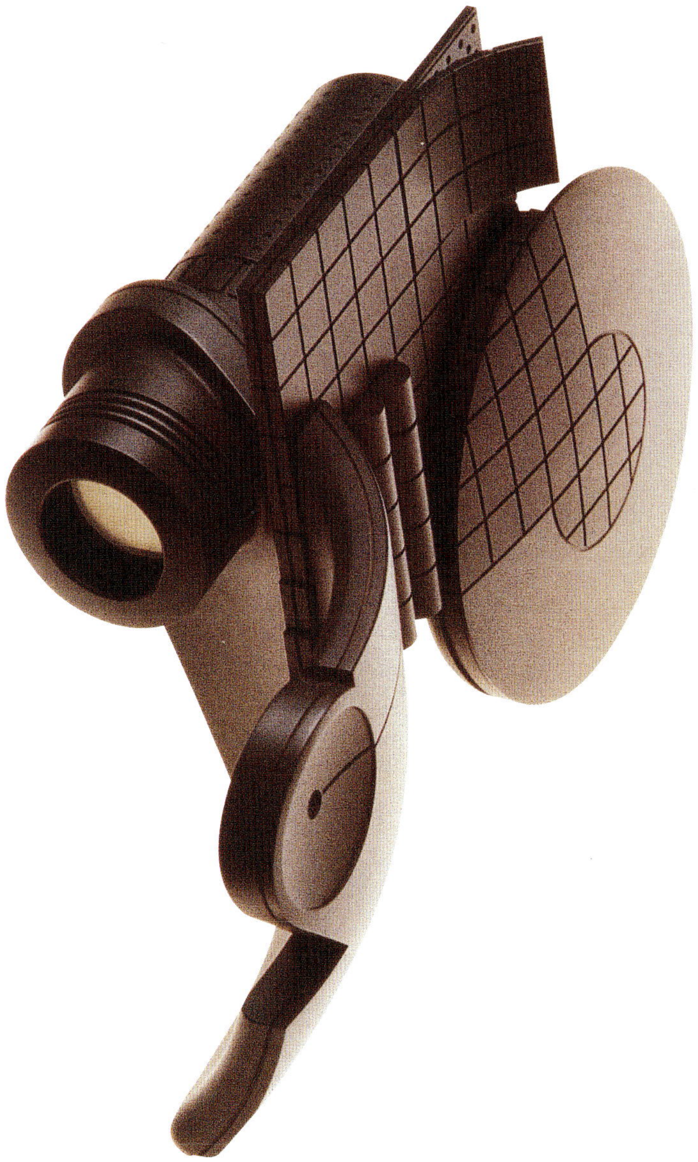

DESIGN LOGIC (DAVID GRESHAM, MARTIN THALER), PROJECT FOR A VIDEO CAMERA, UNITED STATES, 1986.

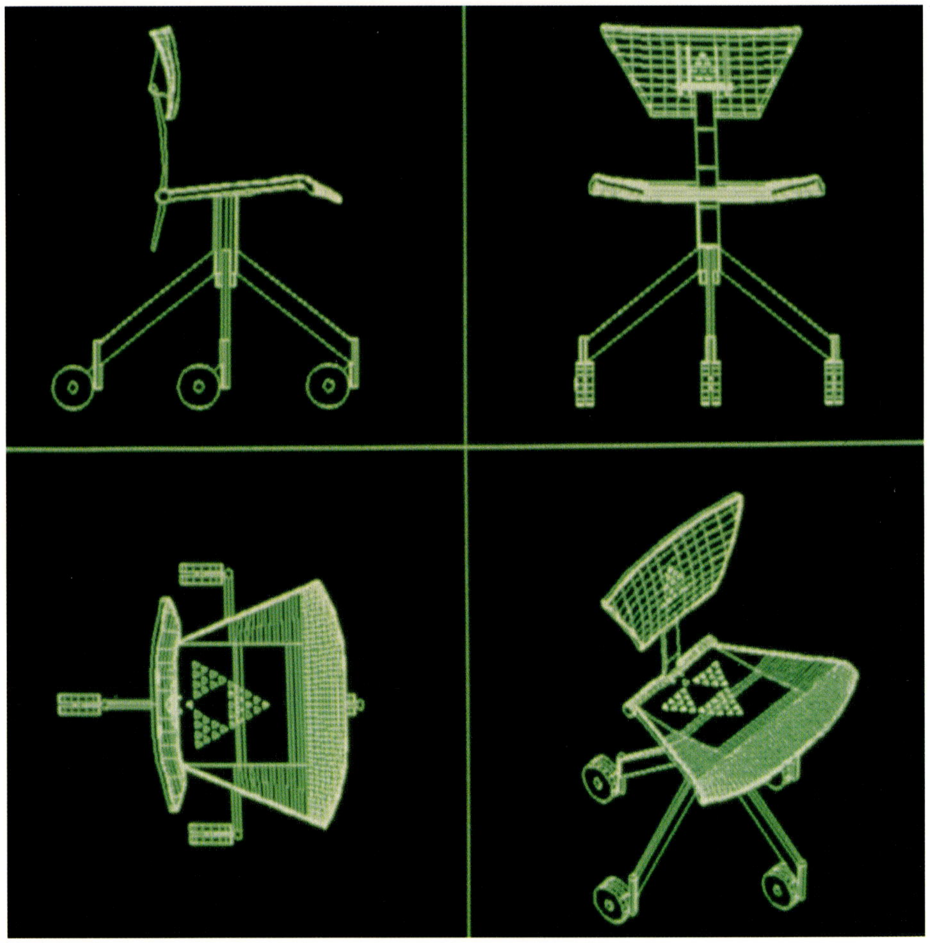

MARC BERTHIER, CAD STUDY FOR AN OFFICE CHAIR, FRANCE, 1987.

produced, have to conform to the exigencies of the "ordinary" manufacturing cycle (to be readily extractable from the mold, for instance), designers instantly grasped that these processes will also enable them to be freed from precisely these constraints.

In France in 2000, at the behest of the CTBA (the Centre Technique du Bois et de l'Ameublement), Anne Gautrot drew up a report entitled "Rapid Prototyping, a New Tool in Furnishing Design." In it, she outlined various processes: stereolithography (3D layering), LOM (Laminated Object Manufacturing), FDM (Fused Deposition Modeling), Selective Laser Sintering of a powdered material coated with a binder, etc.; all means of validating a form by rapidly obtaining a model that allows, in particular, to judge its "feasibility for molding."

The prime goal is to determine speedily whether a project conforms to standard criteria of manufacture. In 2004, with his "Solid" chairs and tables in stereolithography, Patrick Jouin demonstrated that, in the field of furniture as in many others, the intelligent exploitation of a novel technique means making possible today what was impossible only yesterday.

TOWARDS ECOLOGICAL DESIGN

Since the Industrial Revolution, a plethora of new materials have been produced that have widened the scope of manufacturing enormously. In parallel, their means of implementation has developed at such a pace that talk of the man-machine system is redundant, since automation has practically ousted the human from the manufacturing cycle. The golden rule in industrialized countries, where supply often precedes and directs demand, is to produce more to consume more. As there is no ingestion without excretion, the consumer society today is faced by overabundance and by the deleterious effects of its own waste; the relationship between humankind and the environment has become a cause for concern.

When in 1866 the German biologist and naturalist Ernst Haeckel coined the word ecology, he could have had no idea that, by the final quarter of the twentieth century, it would have become both a distress signal and an urgent appeal to help a planet in grave danger. In 1972, at the time of a

MARTIN SZEKELY, L'ARMOIRE, IN ALUCOBOND, CUT OUT AND FOLDED, DESIGNED AND MACHINED ON COMPUTER, 43 1/5 × 25 × 16 4/5 IN. (108 × 64 × 42 CM), MANUFACTURED BY KREO, FRANCE, 1998.

United Nations summit in Stockholm on the theme of "Man and the Environment," certain scientists went so far as to recommend a halt to economic growth. The year 1987 saw the publication of the report "Our Common Future" by the Commission of United Nations, chaired by Mrs. Gro Harlem Brundtland. In it, sustainable development was defined as responding to "the needs of the present without compromising the capacity of the future generations to answer theirs."

By the time of the United Nations conference on the environment and development that took place in Rio in 1992, the adopted convention states, among its general principles, the following: precaution, which consists in stemming environmental pollution; the "polluter pays," which stipulates that polluters must assume the cost of the damage caused by their activities; solidarity, which decrees that environmental protection and the fight against poverty concern all nations, with the responsibilities (even if variously geared) being shouldered together. Sustainable development is also one of the paramount objectives of the European Union. In June 2001, at the time of the Gothenburg Summit, the Commission considered the adoption of a system of production stipulating reductions in the mass of waste, and supporting recovery and recycling.

Recuperation and recycling are not in themselves innovations; since antiquity broken glass was collected for cullet and bronze helmets and weapons were taken from the battlefield to be melted down. At all times throughout history, recovery and reuse of materials has been practiced on both small and industrial scales. Using scrap iron, with the eventual addition of a little cast iron, quality steel has been produced in open-hearth furnaces since 1865. Since 1900, specialized high-strength steels have been exclusively made from scrap in electric-arc furnaces. Half of the world's steel originates in scrap. Be it metals, newsprint and paperboard, glass, textiles, plastics, etc., collecting and recycling have become obligatory throughout the world. Experience proves that it is almost always more economical in terms of energy and less polluting to recycle than to extract raw material afresh.

preceding pages and facing page
IMAGES FROM *DÉCOUVRE LE DESIGN DE MARTIN SZEKELY* BY NADINE COLENO, ÉDITIONS DU REGARD, FRANCE, 2004.

However, certain substances do pose genuine problems. Recycling thermoplastic synthetic materials by melting them down is not as simple as it may appear, but the stumbling blocks do not seem insurmountable. This is not the case with thermosets that attain their final state during a rigorously irreversible chemical reaction. As for the present, the use of energy resources drawn from the calcination of plastics, for want of anything

PATRICK JOUIN, MANUFACTURE OF CHAIRS AND TABLE FROM THE SOLID SERIES, MADE USING STEREOLITHOGRAPHY, FRANCE, 2004.

facing page

top

DETAIL OF A RADIOLARIAN SKELETON.

bottom

DETAIL OF THE SOLID SERIES T ABLE IN STEREOLITHOGRAPHY, BY PATRICK JOUIN.

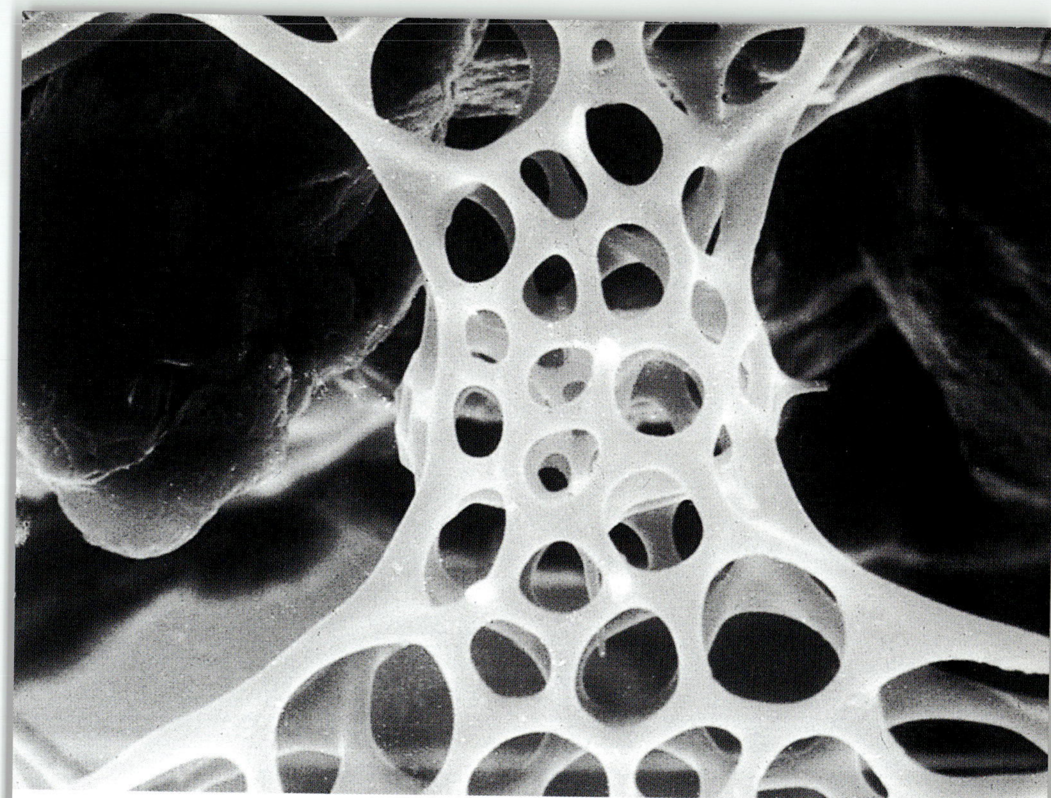

better, has already proven profitable. Derived from oil, coal, or natural gas, plastics are eminently combustible. Breaking down materials in this manner is obviously unsatisfactory both intellectually and economically. To successfully optimize outcomes and tackle what is a complex problem, the most judicious approach is to design any given appliance with a limited number of materials, thus simplifying selective recycling and making it easy to dismantle once out of service. Today, a product cannot be designed without considering how it will end its life.

The principle of "leasing," already applied in certain companies, makes producers responsible for their products during their life span and even beyond. Manufacturers are the owners and thus remain responsible for dealing with a product when it falls out of commission. This constraint obliges them to make provision for the reuse of the entire piece, or of parts when replacing defective components, reconditioning and, if redeployment proves impracticable, recycling the materials.

For the future of the planet, the stakes are high. Yet we can be sure that designers will turn these demanding new working specifications to good account, since minds accustomed to exercising their creativity to the maximum find few things more stimulating than a challenge.

facing page and following pages

PATRICK JOUIN, CHAIRS AND TABLE FROM THE SOLID SERIES, MADE USING STEREOLITHOGRAPHY, FINISHED ARTICLE AND DETAIL, FRANCE, 2004.

FROM MATERIALS TO DESIGN

JEAN GRENIER

Humankind has a wonderful propensity for imagination and creativity.

Sculptors need materials—stone, plaster, or bronze—to express their art. When a designer faces the constraints of a project, whether imposed by a client or self-imposed, he or she does so armed with the complete range of materials known to man.

A distinction can be made between two categories of designer. Independent practitioners tend to work on specific commissions or follow their own personal, non-conformist solutions to a particular problem; they then release their work as one of a kind or in a limited series.

The other, much larger, category comprises both independent and in-house designers who create products intended for mass production. Their most important concern is finding the best shapes to suit the projected usage, the economic constraints, and the manufacturing criteria for production on an industrial scale.

In the former category, economic considerations are distinctly less important than the originality of the work: a free choice of materials is essential when it is to underpin a particular message. This approach also implies a high degree of selectivity, to enable the chosen material to act at the height of its expressive ability or at the limits of its specific characteristics. This choice is guided by considerations of workability and the means of transformation available to a limited output, as well as, of course, the appearance of the finished product.

In the second case, the criteria for the selection of materials will depend on the size of the intended run, the means of transformation available within the relevant industry, the intended usage, and the market's reaction to the changing behavior of the buying public.

What, then, of the materials themselves? This chapter will concentrate on those that have been most widely used in both industrial and small-scale contexts, and those whose future looks bright. It will look at some of the processes involved in working with these materials, the choices to be made, and their influence on manufacturing costs. It will also address technology transfer, the evolution of industrial tools, and the development of special alloys, including those with a shape memory.

preceding page
DRILL BITS

facing page and page 93
SOLID WOOD IS STILL USED TO REPLICATE THE GRANDER STYLES OF THE PAST, BUT REPRODUCTION FURNITURE IS NOW CARVED BY MACHINE FOR UNIFORMITY AND SPEED.

All materials can be broadly divided into three categories:

Natural Materials

Natural Materials can, in turn, be separated into three groups:

- Mineral (e.g., metals, metal oxides, clay, sand)
- Vegetable (e.g., wood, natural fibers)
- Animal (e.g., leather, horn, ivory)

Synthetic Materials

Synthetics can be categorized by the nature of their origin:

- Mineral (e.g., coal, petroleum, silica)
- Vegetable (e.g., cellulose, latex)
- Animal (e.g., milk)

Composite Materials

Composites use both natural and synthetic materials, combining the binding matrix of one with the strengthening properties of the other.

NATURAL MATERIALS

WOOD

For almost as long as humans have walked the Earth, wood has been used for shelter, weapons, and crossing water, for building, sculpting, cooking, heating, and melting other materials. For millennia, although wood was a universally adopted material, the way it was worked evolved very little: it was cut, carved, smoothed, or drilled, and the resulting forms were usually based on solid pieces. With the advent of the first machines, it became much easier to produce thin sheets and strips of wood. Veneering and marquetry became widespread.

The next evolutionary stage did not occur until the twentieth century. A new generation of materials appeared: plywood, reconstituted woods such as chipboard, derivatives made from wood fibers, such as MDF (medium-density fiberboard), and the composite boards that so rapidly became widely used in furniture and shelving, where metal or PVC facing is bonded onto MDF.

What does the future have in store for wood? One possible answer lies in the field of insulation. Wood fiber is a popular choice for sound insulation, and wood is also one of the best thermal insulators. Ordinary balsa wood

MARQUETRY. TONAL CONTRAST IN THE SELECTION OF THE WOOD ALLOWS FOR COMPLEX, DETAILED COMPOSITIONS, SUCH AS THIS ALLEGORICAL ARRANGEMENT OF MUSICAL INSTRUMENTS.

has exceptional properties of insulation, and is used, for example, in reheating compartments in hospital meal carts.

The mechanical properties of wood are no less valuable. To take the example of balsa again, while it is very fragile along the grain, it has exceptional strength on its end grain compared to its density. It is used in a sandwich configuration in the construction of aircraft water tanks, with capacities of tens or even hundreds of thousands of gallons.

When wood is sliced and recombined as glued laminated timber, or "glulam," it can make lightweight supporting structures to span wide areas, such as the roofs of gymnasiums, factories, and other large buildings. Some of the very dense woods available have been found to be very effective fire retardants.

Finally, wood can be chemically treated, either by surface application or impregnation under pressure, to make it fireproof and waterproof, to prevent the spread of insects and rot, to change its color, to harden it, or to improve its mechanical properties.

Primary processing makes wood available in various forms, both in its "raw" state, as timber for joinery, furniture, or packing, or in the form of derived materials, such as particleboard and plywood. Wood's various characteristics and appearances have allowed a great number of uses throughout the human environment.

One of the reasons why wood has been used by sculptors all over the world is that it can easily be shaped into complex curved surfaces. This is equally important in an industrial context, to the model maker, and to the prototype pattern maker.

Some exceptionally beautiful forms have been achieved with wood over the centuries—from cathedral arches built around wooden frames to the hulls of boats, and from stagecoaches and carriages to automobiles and airplanes.

And what of musical instruments? The lute, the harp, the cello, the double bass, and the violin, to name but a few from the western world, are constructed to express functional, pure, and ergonomic forms, while producing complex sounds. The way an instrument resonates is a matter of complex alchemy, depending on the density of the materials used, the nature of the structures involved, and the instrument's shape, dimensions, and final mass.

facing page and below

MARTIN SZEKELY, *COROLLE,*

PYLON FOR HIGH-VOLTAGE ELECTRICITY LINES, GLUE-LAMINATED SCOTCH PINE (HEIGHT 100 FT), FRANCE, 1998.

following pages

CONSTRUCTION OF A BUILDING WITH A LARGE, OPEN AREA AT GROUND LEVEL. RADIATING BEAMS PREFABRICATED OUT OF GLUE-LAMINATED TIMBER.

BAMBOO SCAFFOLDING,

HONG KONG, 1995.

One can imagine what a designer must feel when asked to work with wood: keen to create and innovate with this sensitive material, ready to bring into play personal and cultural sensitivities and technical knowledge. Yet the attitude is not the same for a mass-produced item as for a limited series or one-of-a-kind piece. From a common wood one demands a maximum of efficiency, while from a precious wood one aims to extract the maximum expression, as when choosing ebony, rosewood, amaranth, or sandalwood for a pen, desk accessory, piece of jewelry, fashion accessory, or watchcase.

Bamboo must be counted among the different types of wood, with its typical wheat-stalk hollow structure and very high inertia for its weight. The most spectacular examples of its use are to be found in the scaffolding on skyscrapers in Hong Kong or Shanghai, which sometimes reaches the twentieth floor. Bamboo offers an ideal combination of lightness, flexibility, and sturdiness. It is composed of very long, unidirectional fibers, and has very similar characteristics to the carbon fiber sheets and strands used in composite materials for high-tech applications.

But, while solid wood is still in common use today, its derivatives are ever more present in our everyday environment, both in furniture and fittings, and in buildings. Derivatives—in the form of boards—are perfect for large-scale manufacturing because of their statistically consistent size and quality, allowing them to be cut without fear of encountering defects. Plywood is one of the most common of these.

Plywood was developed from the thin leaves of wood used for veneering. Sheets of plywood are obtained by transverse or rotary slicing from the timber log, stuck together, with the grain running in alternate directions from one layer to the next. The resulting board is free from the flaws and dimensional constraints of an ordinary timber plank, and its mechanical resistance to a given load is almost identical in all directions. The idea of producing plywood on an industrial scale was first mooted in the second half of the nineteenth century. Production began in the early twentieth century, and was boosted during World War I, when large quantities were needed to mass-produce fighter plane cockpits.

Then, when architects started to visualize buildings based on a huge concrete shell, easy-to-bend plywood turned out to be the ideal material for shuttering. Some of the more spectacular examples of this are the Sydney Opera House and the CNIT building at La Défense in Paris.

following pages
CHARLOTTE PERRIAND, CHAISE LONGUE WITH WOODEN BASE AND WOOD AND BAMBOO SEAT, JAPAN, 1940. ADAPTED FROM THE STEEL TUBE-FRAMED CHAISE LONGUE LC4 BY LE CORBUSIER, JEANNERET, AND PERRIAND, 1928.

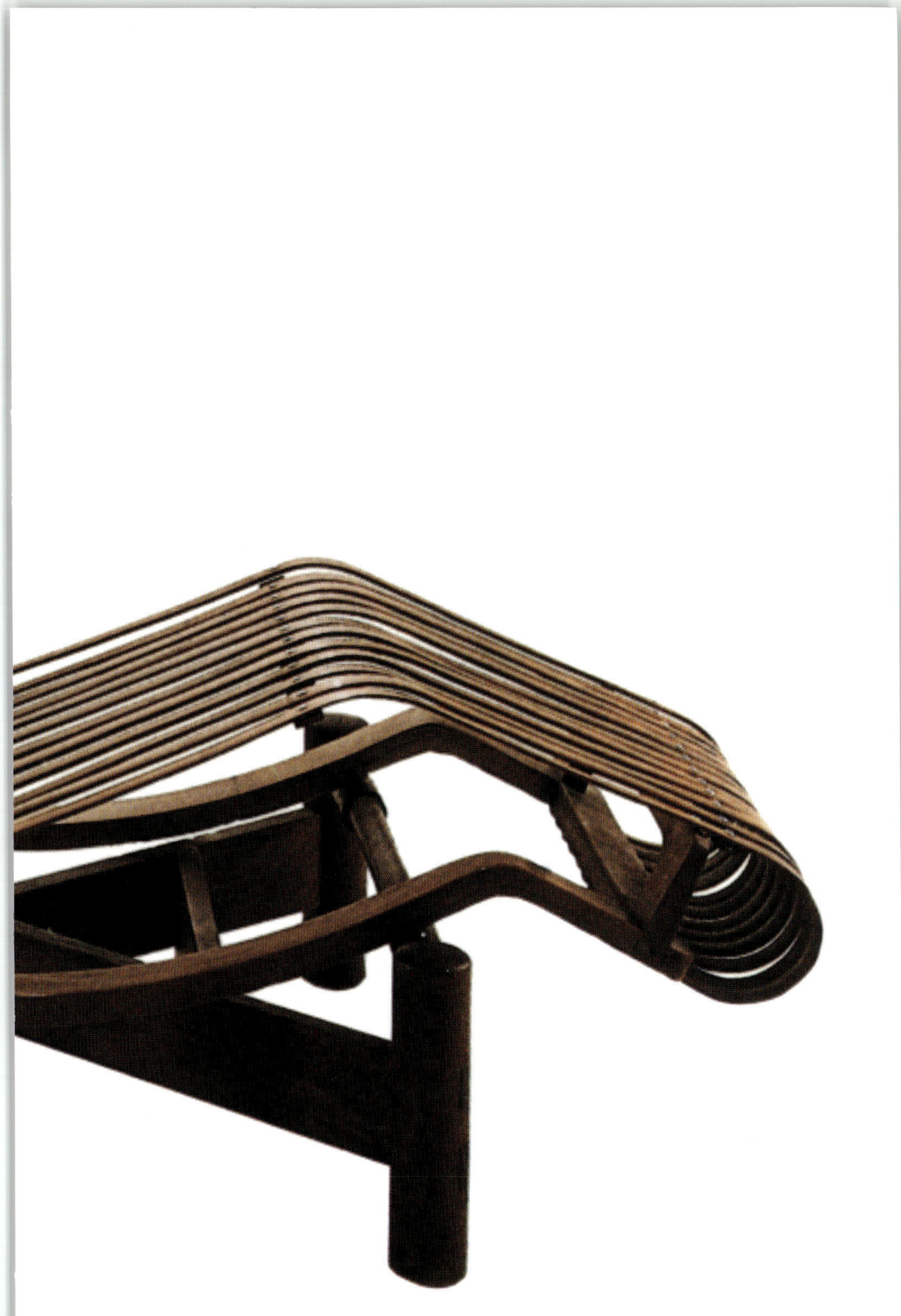

CHARLOTTE PERRIAND, STUDY FOR SOLID WOODEN DESK WITH THREE LEGS, FRANCE, 1938.

facing page
CHARLOTTE PERRIAND, FREE-FORM TABLE FOR MOUNTAIN CHALET, SAINT-NICOLAS-DE-VÉROCE, HAUTE-SAVOIE, 1938.

Blockboard is based on a similar principle to plywood. It is not an accumulation of layers like plywood, but instead is a thick core of wood strips glued together, sandwiched between layers of veneer front and back.

Sheets of plywood and blockboard were produced industrially from around 1911, but it was not until the 1930s that particleboard began to be manufactured. This is reconstituted from wood shavings or chips, stuck together by thermosetting resin, which polymerizes at high pressure and temperature. Particleboard has been used extensively in architecture and furniture, being adapted in the latter case by the addition of a wood veneer or metal facing. An especially popular early development was the addition of a decorative laminate facing made from layers of brown paper impregnated with a phenol-formaldehyde resin, the decorative top sheet being protected by a layer of transparent melamine. The possibility of volume-dying melamine, unlike Bakelite, meant that it could be used to give the panel a uniform color.

Fiberboard is made from wood or flax fibers, which are broken down under steam pressure, then bound by thermosetting synthetic resins (urea-formaldehyde, melamine, or phenol-formaldehyde). Today, medium-density fiberboard, or MDF, can be produced with exceptional toughness, and considerable thickness. Given the ease with which it can be worked, it is replacing solid wood in a great many applications. It is much appreciated by numerous designers: Martin Szekely's Container series of 1987 is one influential example.

The last of these composite boards is hardboard. It is created through a method known as the "wet process," in which wood fibers are bonded together under extreme heat and pressure, with no glue added. Hardboard has similar characteristics to its distant cousin, cardboard, in that it does not like a damp environment. It is still used extensively in furniture when mass distribution—which is to say, low cost—is a prime consideration. All these boards are manufactured relatively cheaply, and allow a high level of regularity, whatever their specific technical characteristics.

This chapter would not be complete without mentioning the technique of bending wood for use in furniture. The thermoformability of wood—i.e., its capacity to adapt its shape under the action of heat—is a particularly useful characteristic in many technical applications. The age-old technique of shaping solid wood by the combined

preceding pages

THE MANUFACTURE OF PLYWOOD

left, top

DEBARKED LOGS, READY FOR ROTARY CUTTING.

left, bottom

ONCE STEAMED, THE LOG IS PLACED ON A VENEER-PEELING LATHE, WHICH WORKS LIKE A PENCIL SHARPENER, REMOVING A CONTINUOUS SHAVING UP TO TEN FEET WIDE.

right, top

A ROLL OF VENEER OBTAINED FROM ROTARY CUTTING OF THE LOG, READY TO BE CUT TO APPROPRIATE SIZES FOR PLYWOOD SHEETS. AFTER DRYING, THE LAYERS OF VENEER ARE GLUED AND ASSEMBLED IN STACKS TO THE REQUIRED THICKNESS.

right, bottom

THE RESULTING STACK IS PRESSED BETWEEN HEATED PLATES, AND SUBJECTED TO A PRESSURE OF 170–215PSI FOR FIVE TO TEN MINUTES. THE GLUE IS POLYMERIZED BY HEAT: THE TEMPERATURE VARIES DEPENDING ON THE TYPE OF GLUE USED AND THE THICKNESS OF THE PANELS.

PLYWOOD SHUTTERING FOR THE PALAIS DU CNIT (CENTRE DES NOUVELLES INDUSTRIES ET TECHNOLOGIES), LA DÉFENSE, PARIS, 1957.

below

PARTICLEBOARD PIANO, FRANCE, 1950S.

facing page

FURNITURE AND CUPBOARD DOORS FACED WITH DECORATIVE LAMINATE (FORMICA), ITALY, 1956.

action of heat and wetting steam was brought back into vogue by Michel Thonet in the middle of the nineteenth century. Scandinavian designers pursued the technique in the 1930s and 1940s, as seen in Bruno Mathsson's Eva and Pernilla armchairs in Sweden in 1934, and Hans Wegner's Peacock armchair in Denmark in 1947. But the technique most used since the end of the nineteenth century involves shaping stacks of thin, completely flexible strips around a template. Glue is applied between the layers, resulting in a rigid and permanent curve. The famous Thonet bentwood chair was made in this way in around 1880, the result of experiments that had begun in 1836. In the early 1930s, Alvar Aalto used laminated, glued wood systematically for load-bearing parts of his furniture. In the 1940s and 1950s Carlo Mollino produced highly experimental designs from molded plywood in Italy, while Charles and Ray Eames in the United States from 1945, and Arne Jacobsen in Denmark from the beginning of the 1950s, all did wonderful things with the material. And the trend continues, seen in work such as Ron Arad's 1993 Empty Chair and Frank O. Gehry's remarkable Powerplay chair from 1990, made from interwoven bent plywood strips.

With plywood as well as with glue-laminated timber, the mineral and organic glues of the past have been replaced today by thermosetting resins. Once polymerized, they maintain the shape they are given. The same principle is used in making oriented strandboard, or OSB.

The influence of heat is also of great importance when it comes to wood derivatives. With a combination of wood and polypropylene (PP) fibers in the fiber mat, a board can be molded straight into a given shape by thermocompression. Some particle and fiberboard flat panels can also be postformed by thermocompression. The thermoformability of the material is vastly improved by the inclusion of fibers or layers of film, made of a thermoplastic polymer.

A process that also uses high-temperature resins, and is currently undergoing rapid development, is the extrusion, and even injection molding, of a mixture of wood and plastic. The molding material can contain up to 60 percent wood fibers or wood flour.

Wood is composed of around 50 percent cellulose, making it a primary source for a major component of synthetic materials, cellulose acetate, otherwise known as celluloid. This will be examined more closely in the chapter on synthetic materials. Synthetics, from the point of view of

facing page
MICHAEL THONET,
SOLID BENTWOOD CHAIR,
VIENNA, C. 1850.

following pages
left
TOOLING PLAN FOR THE GLUED LAMINATED WOODEN CHAIR SHOWN ON FACING PAGE.

right
MICHAEL THONET, GLUED LAMINATED WOODEN CHAIR, BOPPARD AM RHEIN, GERMANY, 1934.

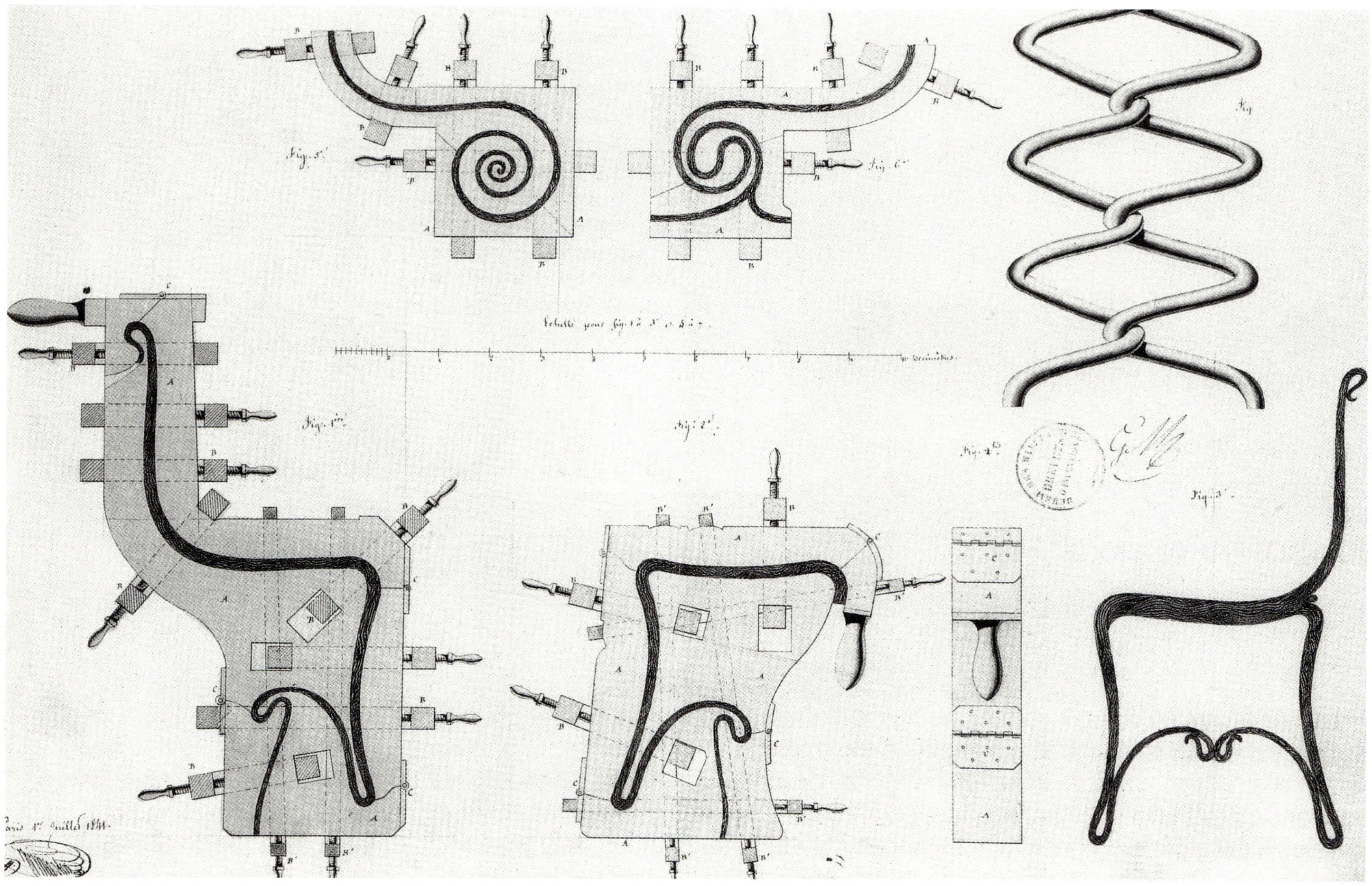
Fig. 5.
Fig. 6.
Fig. 1.
Fig. 2.
Fig. 3.

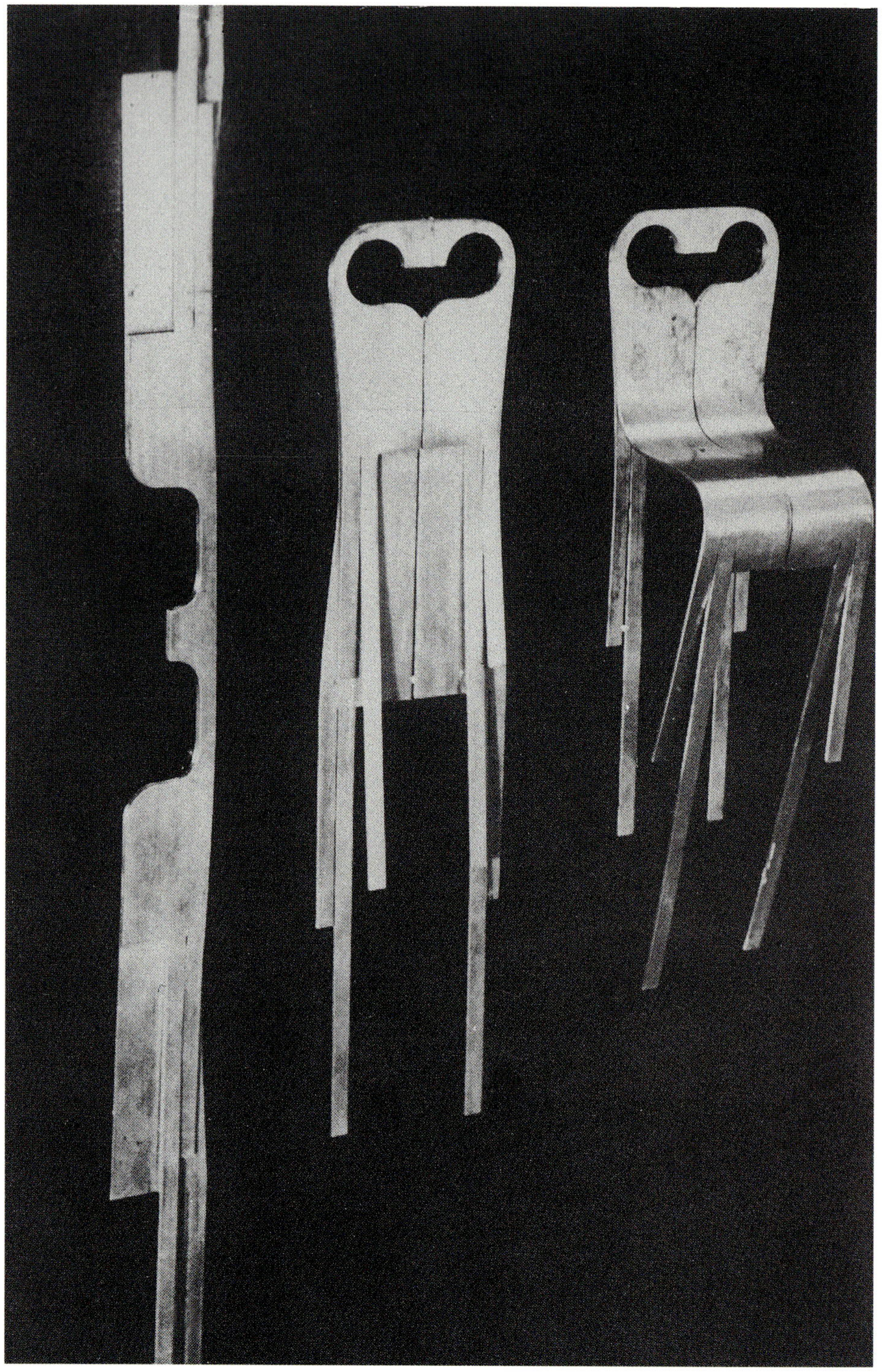

preceding pages

left

THE LAYERS OF PLYWOOD NEEDED TO MAKE THE CHAIR SHOWN ON THE FACING PAGE, PRE-CUT INTO STRIPS.

right

FRET-CUT PLYWOOD CHAIR, **THONET STUDIOS**, C. 1880. STUDIES FOR THIS CHAIR ARE BELIEVED TO HAVE STARTED UNDER THE SUPERVISION OF MICHAEL THONET IN 1860.

facing page

BRUNO MATHSON, PERNILLA ARMCHAIR WITH FOOTREST, PRODUCED BY KARL MATHSON, SWEDEN, 1934.

below

CARLO MOLLINO, ONE-OFF DESK FOR THE INSTITUTO DI COOPERAZIONE SANITARIA IN TURIN, TODAY PART OF THE DESIGN COLLECTION OF THE POMPIDOU CENTER. MADE FROM MOLDED PLYWOOD BY THE APELLI E VARESIO STUDIOS IN MILAN, 1950.

FRANK O. GEHRY, POWERPLAY
CHAIR AND OTTOMAN,
ISSUED BY KNOLL, USA, 1992.

facing page
BATCH MOLDING OF CURVED
SHAPES IN PLYWOOD.

"3D-VENEER," THIN PLYWOOD THAT CAN BE SHAPED WITHOUT THE WOOD (BEECH, MAPLE, OAK, WALNUT) CRACKING OR SPLITTING. MADE BY REHOLZ, GERMANY.

facing page
FASAL, A RAW MATERIAL IN PELLET FORM MADE FROM RENEWABLE MATERIALS SUCH AS WOOD AND MAIZE, MIXED WITH NATURAL RESINS AND PLASTIFIERS. WHEN HEATED, IT CAN BE INJECTION-MOLDED OR EXTRUDED WITH THE CONVENTIONAL MACHINERY USED FOR PLASTICS. MANUFACTURED BY AUSTEL, AUSTRIA.

organic chemistry, are also of vegetable origin: micro-organisms and plankton accumulating at the bottom of the oceans to form organic layers between sedimentary rock and eventually turning into oil and hydrocarbons. In our time we have learned to put this prehistoric transformation of living tissue to good use, using intelligent and complex processes to invent new, "plastic" materials.

LEATHER

Leather is another organic material. The generic term "leather" covers mammal skins of all sorts, including hide from horses and cattle, and skins from goats, sheep, and lambs.

For millennia, animal hides of all shapes and sizes have been widely used by humans. In prehistoric times, simple skins were used as clothing. As tanning techniques were developed, the conservation of leather was improved and its uses gradually extended to a number of everyday items, from the thoroughly utilitarian (shoes and clothes; horse tack) to the ornamental and artistic. In more recent times, it has been used for bags, purses, wallets, desk accessories, and for covering furniture and car seats. Designers have largely concentrated on bags, classic travel goods, and of course furniture, sometimes using leather to surround metal structures. Some of the desks designed by Jacques Adnet at the beginning of the 1950s come to mind, as does Mario Bellini's Cab chair for Cassina in 1976. The centerpiece of Philippe Starck's Richard III aluminum-frame chair, in its original 1982 version made for the French president François Mitterrand's private apartments at the Élysée Palace, is a black leather seat cushion.

The quality of a piece of leather depends to a great extent on the condition of the original skin. If the animal has lived in an enclosure surrounded with barbed wire, it is likely that the skin will have been scarred. Some may even have had botflies nesting under the skin, leaving behind open holes. Sometimes a cow or ewe will have had a caesarian section when giving birth, and the large scar will make the usable part of the hide smaller and more expensive. Chemical tanning can help here by adding surface layers, which modify the properties of the leather and mask defects. These layers are formed either by impregnating the surface with a polymer, in the form of a paint or varnish, or by transfer of a PVC plastisol coating.

BELLY LEATHER, LEATHER MADE FROM A COW'S RETICULUM. THIS FASCINATING HIDE IS TANNED UNDER ECOLOGICALLY SOUND CONDITIONS AND ITS PRODUCTION IS STRICTLY MONITORED. IT IS WATER RESISTANT AND COMES IN A VARIETY OF TEXTURES AND COLORS. THE MANUFACTURER (NATURE TOUCH, BRAZIL) ALSO PRODUCES RUMEN FROM COW'S STOMACHS, AND "BULLFROG" SKIN. THESE SPECIALTY LEATHERS ARE USED FOR BAGS, SHOES, CLOTHING, FASHION ACCESSORIES, AND FOR COVERING FURNITURE.

But the inherent fragility of the natural structure of leather has also prompted engineers to find ways of artificially combining collagen and various natural and synthetic fibers, and combining them mechanically or chemically in layers. The aim is to produce a synthetic leather, which can be cut to large sizes, with identical properties at all points on its surface. Poromerics such as Corfam, and other leather substitutes such as Clarino and Gore-Tex, came out of this innovative research in the 1970s.

The most remarkable of the leather substitutes is Skaï. The trademark was registered by the German surface specialists Hornschuch at the end of the 1950s. The original idea was to create a product that would mimic the appearance of leather, but without its disadvantages and at a much lower cost. Hornschuch's engineers had the idea of coating a knitted "jersey" base fabric with a PVC plastisol coating (PVC in its solvent phase). Within a few years, the product became a global phenomenon. Ever since, in Europe, "skaï" has come to mean almost any coated fabric: the brand name has become a proprietary eponym to rank with Kleenex and Thermos.

In the course of the 1960s and 1970s, Skaï gained many followers in the world of fashion, as it was used extensively for clothes, shoes, and accessories. It was also famously used for the outer covering of the Sacco beanbag chair created by Piero Gatti, Cesare Paolini, and Franco Teodoro for Zanotta in 1968.

facing page
THE TANNERY AT METZ, NORTHEASTERN FRANCE, IN THE EARLY TWENTIETH CENTURY.

following pages
FLESHING SKINS AT THE BRÈDE TANNERY, GIRONDE, FRANCE, 1927.

above, facing page and following page
FRANÇOIS AZAMBOURG,
ARMCHAIR SEAT MADE FROM A SANDWICH OF LEATHER AND POLYURETHANE FOAM, MOUNTED ON A BASE ASSEMBLY OF WELDED STEEL RODS. HERMÈS, DOMEAU & PÈRES, FRANCE, 2005.
THE CONSTRUCTION OF A REDUCED-SCALE MODEL AND MANUFACTURE OF THE SEAT.

MINERALS

ROCKS

Rock is what the earth's crust is made of. All rock falls into one of three groups, depending on its origin: sedimentary, igneous, or metamorphic. Alternatively, we can categorize minerals by their physical characteristics: whether they are homogenous, hard, and potentially very close grained like quartz, or malleable like clay, or soft like sand, or even liquid like petroleum. Among the hard rocks, marble and granite are most likely to interest the designer. Their use in construction, sculpture and the crafting of everyday objects stretches back beyond record, as is clear from the numerous artifacts handed down from earliest prehistoric times. Granite and marble can easily suggest a feeling of the sacred and symbolic by association with their use for religious edifices and funeral monuments.

The weight of these materials, the difficulty of extracting them, and the slow rate of working with them, added to the delicate handling needed when transporting them due to their fragility, have meant that large-sized projects have always been problematic. But when it comes to design, large has rarely been the rule. In the 1960s, designers like Angelo Mangiarotti were instrumental in the renaissance of interest in the use of stone and marble in furniture and furnishings. Mangiarotti's Eros stools and tables (1969) made of turned, cut, and polished marble opened the gates to other designers: Lodovico Acerbis and Giotto Stoppino, Andrea Branzi, Ettore Sottsass, and Enzo Mari. Michele De Lucchi's Kampur table, created in 1988 for Up & Up, used a particularly avant-garde technique. The top was designed as an inlay of white Carrara marble and black marble from Marquinia in Spain. A computer-controlled water-jet machine was used for the cutting; the shapes extracted from the white marble were memorized by the computer, and cut out of the black marble with the same precision. The assembled white-on-black or black-on-white design (both negative and positive versions were sold), lent itself to production in larger numbers. The table stem was in black marble.

preceding page

PHILIPPE STARCK, RICHARD III ARMCHAIR, CREATED FOR PRESIDENT FRANÇOIS MITTERRAND'S APARTMENTS AT THE ELYSÉE PALACE. ALUMINUM STRUCTURE PARTLY COVERED WITH LEATHER, 1982.

ANGELO MANGIAROTTI,

EROS TABLES AND STOOLS,

TURNED AND CONTOUR-CUT

MARBLE, ITALY, 1969.

above

CUTTING A BLOCK OF STONE.

right and facing page

MICHELE DE LUCCHI,

CONSTRUCTION OF KAMPUR TABLE,

CARRARA AND MARQUINIA MARBLE,

PRODUCED BY UP & UP, ITALY,

1998.

following pages

left

ENZO MARI, GREY MARBLE

ASHTRAY, ISSUED BY DANESE,

ITALY, 1964.

right

ENZO MARI, WHITE MARBLE TABLE

ORNAMENT, ISSUED BY DANESE,

ITALY, 1964.

CERAMICS

Some historians date man's first use of clay to the beginning of the Mesolithic era, around ten thousand years ago. At first, clay was simply dried in the sun. For containers to be usable, the clay had to be added to a rigid framework (wicker for example) or reinforced with a fibrous material like straw, to increase the tensile strength. Thus strengthened, clay became a distant ancestor of today's composite materials. Dried clay was also carved into votive sculptures, body ornaments, and such.

The first bricks—forerunners of the standardization of components used in construction ever since—were also unfired. To this day, in hotter parts of Africa and Asia, sun-dried clay bricks are still used.

Then came the age of fired clay. First, clay was fired directly on the ground, in low furnaces reaching temperatures of around 1100°F. By 3000 B.C.E, the first kilns were being used in Syria. Thus began a long period of constant development, from ancient times to the present day. The evolution of ceramics is dotted with masterpieces, as decorative techniques succeeded each other, such as applying engobe (liquid slip), and, in the first millennium, glaze. These techniques have been used to create artifacts in almost every culture. Porcelain appeared in China during the T'ang dynasty (around 650 C.E.).

The list of earthenware masterpieces that have been created over the centuries is endless. Focusing on just a small part of its history, and looking, for example, at the period between the late nineteenth-century Arts and Crafts movement in Great Britain to the present day, there is evidence of a wealth of activity. To cite the ceramic tiles of William Morris (1870) and the contemporary work of the Memphis group and Matteo Thun's strange vases and tea and coffee services (1981) leaves, of course, an impressive number of major works unmentioned.

Some designers have held onto the traditions of the potter's craft: the postwar years were marked by a considerable influence from the Japanese master Shoji Hamada, who—like the British potter Bernard Leach—was a devotee of the Mingei (folk art) movement as made popular by Yanagi Soetsu. Others have taken mass-produced ceramics to a high level of creativity, such as with the Liekki service designed by the Finnish potter Ulla Procopé for Arabia in 1957, or the TC stackable tableware for hotels made

preceding pages

BASALT FAB, A CLOTH MADE FROM CONTINUOUS FILAMENTS OF BASALT, WITH A MELTING POINT OF 2,650° F. THIS INEXPENSIVE MATERIAL IS USED FOR FIREPROOFING BUILDINGS AND AIRCRAFT AND INSULATING INDUSTRIAL FURNACES, AND IS A COMPONENT OF COMPOSITE MATERIALS IN THE AUTOMOBILE INDUSTRY. MANUFACTURED BY THE MASUREEL GROUP, BELGIUM.

following pages

left

GIO PONTI, PORCELAIN WASHBASIN FOR IDEAL STANDARD, ITALY, 1953.

right

RAYMOND LOEWY, 2000 COFFEE SET FOR ROSENTHAL, GERMANY, 1954.

CLAY BRICKS DRYING IN THE SUN.

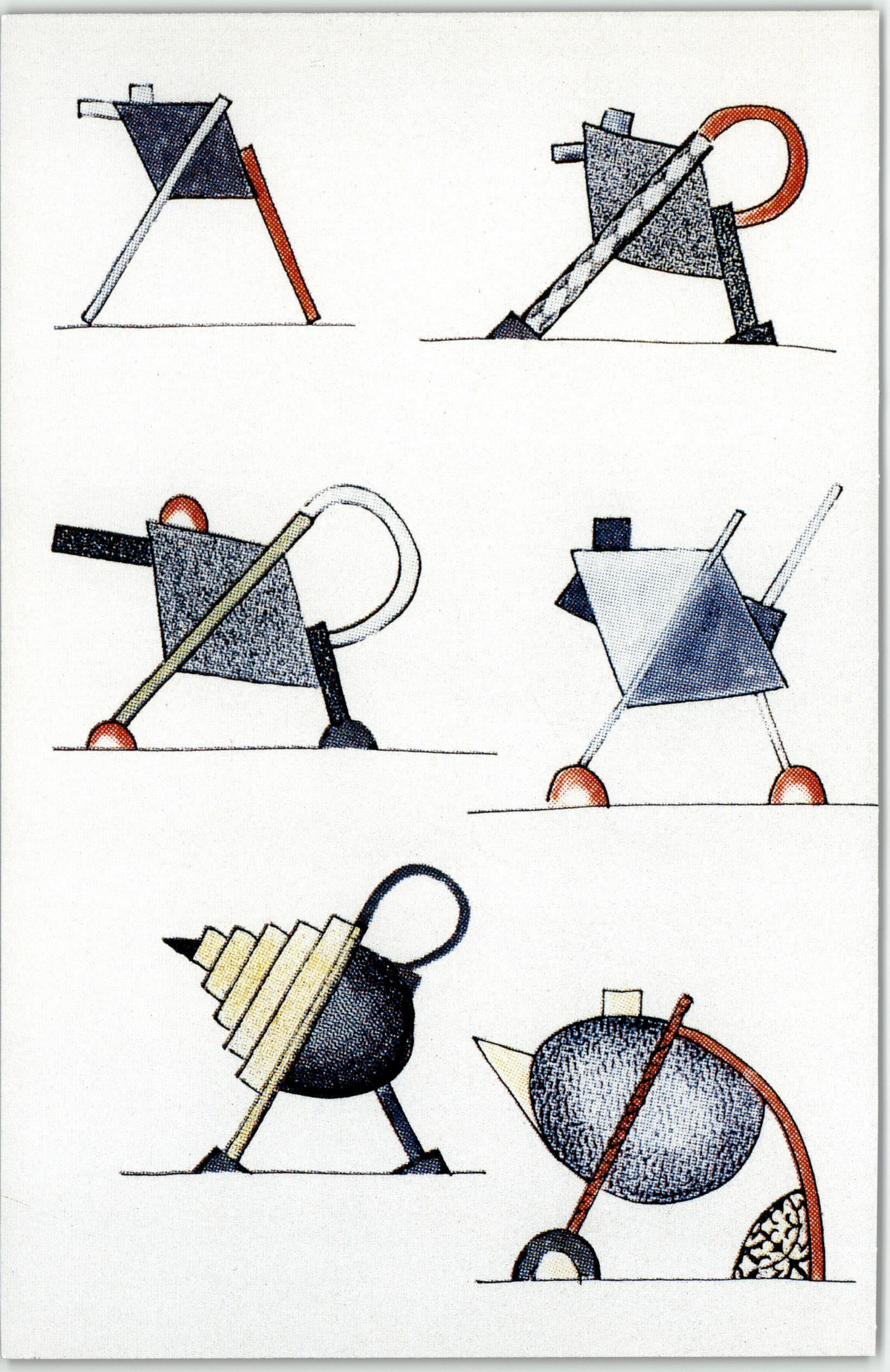

for Thomas/Rosenthal AG by the German designer Hans Roericht in 1958, while he was still a student at the HfG design school at Ulm. A classic example of sanitaryware would be the P series (lavatory, basin, etc.) created by Gio Ponti for Ideal Standard in 1953.

Ceramics are made from materials with a high plasticity, such as kaolin, clay of various types, and colorings, combined with nonplastic materials in a powdered state, such as sand, calcified flint, chamotte, potsherds, cement, and feldspar flux, to which water is added during shaping to form a paste.

Ceramic paste can be volume-dyed, but since a large amount of coloring agent is needed, the process is only used for relatively thin pieces. However, iron oxide, which is already present as finely distributed impurities, can impart a color from yellow-ocher to brownish-red, or even black, given a sufficient concentration and a high firing temperature.

The most commonly used techniques for shaping ceramic pieces are slip casting and jiggering. Casting has a long history, and can be done with basic equipment: it involves filling absorbent molds with a very liquid clay and water mixture called slip. The highly absorbent mold, traditionally made of plaster, absorbs the water from the mixture, and the solid matter left behind forms the final piece. Both solid and hollow shapes can be cast: for hollow casting, the clay suspension is in constant contact with the walls of the mold and the water is absorbed by capillary action, until a sufficient thickness is built up and the remainder is poured out. In some cases the technique of blow molding is used to manufacture large, hollow pieces that might otherwise collapse under their own weight.

A closed mold is used to case solid, thin pieces, such as plates. The two facing plaster cavities absorb water from the slip simultaneously.

Injection molding is also used with ceramics, as a technique specially developed for the mass-production of tableware. As with injecting thermoplastic synthetic resins, the slip is introduced under pressure into a press-mounted mold.

With jiggering, a flat "batt" of clay is laid over a rotating mold, and a template on a lever arm presses it down, forcing the clay onto the mold and applying a profile to the outer surface at the same time.

preceding pages

left

MATTEO THUN, TEAPOT STUDY, ITALY, C. 1980.

right

ETTORE SOTTSASS, ESMERALDA VASE, 1994. NEW PORCELAIN, HEIGHT 18 1/2 IN. MADE AT THE MANUFACTURE NATIONALE DE SÈVRES, FRANCE.

facing page

CERAMIC HIGH-VOLTAGE INSULATORS AT THE COMPAGNIE GÉNÉRALE D'ÉLECTRO-CÉRAMIQUE, ANDANCETTE, FRANCE, AROUND 1970.

MOLDS FOR CERAMIC WARE BEING PREPARED FOR SLIP CASTING, CHINA, 2000.

facing page
AN UNFIRED PORCELAIN COFFEE FILTER IN ITS PLASTER MOLD, CRAFT (CENTRE DE RECHERCHES SUR LES ARTS DU FEU ET DE LA TERRE), LIMOGES, FRANCE, 2005.

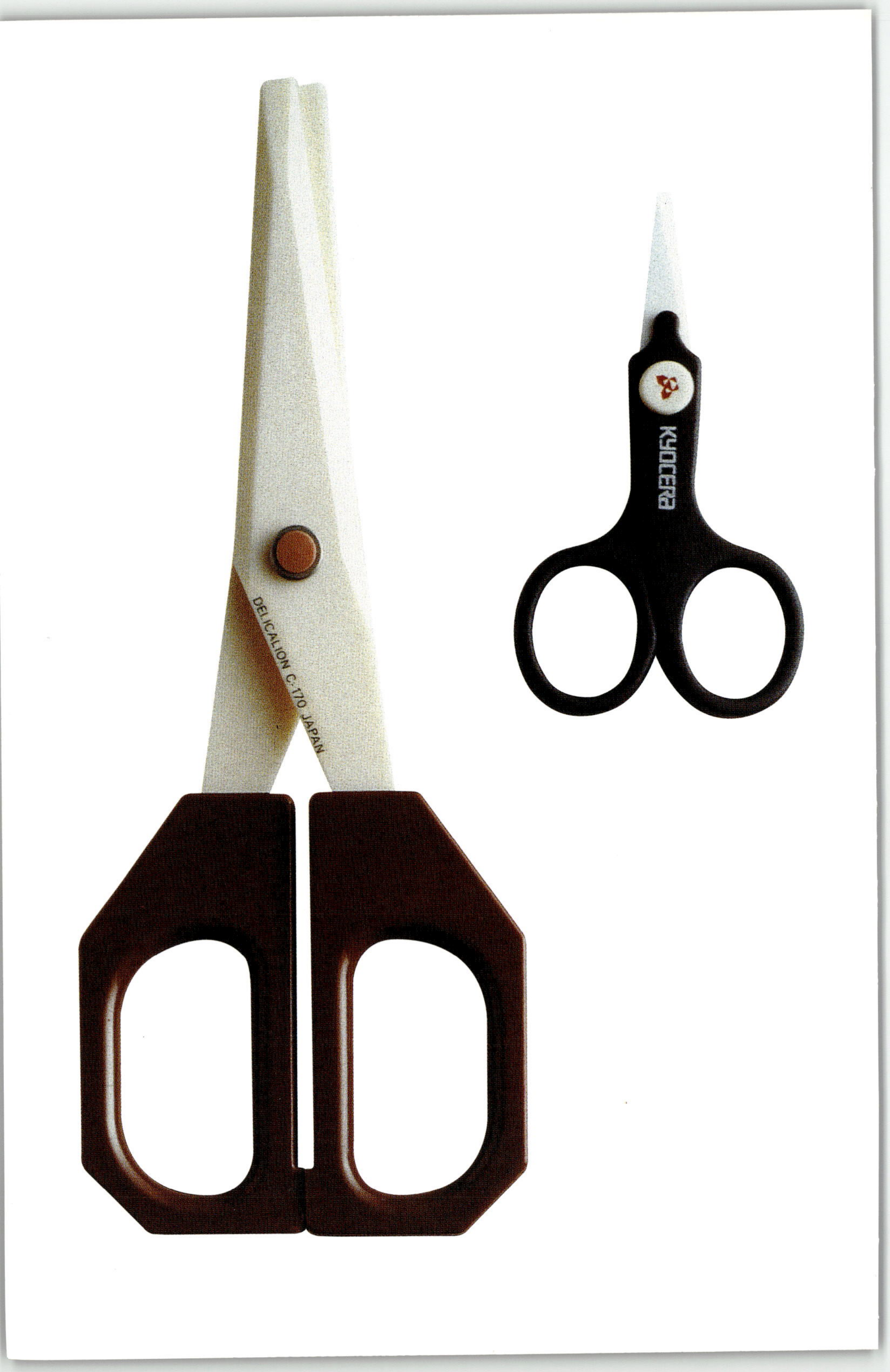
DELICALION C-170 JAPAN
KYOCERA

Malleable clay can also be shaped by pressing. The traditional technique uses a press that simply pushes the mixture into a mold. Isostatic or hydrostatic pressing is a more recent technique, which consists of pressing a granulated mixture with a very low water content (4–8 percent). The powerful presses can apply between 2,000–4,000 psi for clay-rich mixtures, and from around 10,000–20,000 psi for clay-free mixtures. The process uses two methods: "wet-bag" and "dry-bag."

Wet-bag isostatic pressing is used for solid pieces and is relatively slow in operation. The mold is constructed of a steel core lined with a polyurethane sheath. The kaolin mass is shaped first by the movement of the press, then by the application of hydraulic pressure underneath the sheath.

Dry-bag pressing has a higher degree of automation, and a faster output. It is used to manufacture hollow pieces of average depth. The mold itself is semi-flexible with a metal core and is similar to those used for the compression of thermosetting synthetic materials.

Apart from the improved speed of molding offered by isostatic pressing, the rate of production is increased further by the elimination of the drying phase: the products come out of the press and are taken directly to be fired.

After a piece has been given its basic shape, it may need further, more delicate work. It is much easier to make holes in clay when it is "green" than when it is dry, but care has to be taken not to misshape the object. Similarly, it is easy enough to cut pieces out of fresh clay, but when high precision is required in the result, this is done on fired pieces, in order to avoid the unknown shrinkage factor during firing. Modern cutting machines can be programmed to follow a design held on a computer, using water-jet or laser-cutting tools.

Extrusion is a process used to manufacture a great number of products. It permits the quick and continuous mass production of basic construction materials. The ceramic paste is pushed through a die, which gives it its shape. As it comes through, a cutting machine is programmed to section it into a continuous series of pieces of a given length. A conveyor belt takes the pieces off for drying and firing. Simple, rough-cut shapes, either closed or open, can be made in this way.

SCISSORS WITH CERAMIC BLADES.

The future looks bright for ceramic materials; they have already become very important in many high-tech industrial applications. Developments in ceramics have been an integral part of some recent and influential technological innovations, which in turn have significant impact on the economic activity of a number of countries.

Specialist research has concentrated on the composition and behavioral properties of ceramics, as well as methods of producing different shapes. Much work has been done on the application of ceramic coatings by thermal spraying. This has many applications in mining and manufacturing industries, and is used for gaskets for airplane engines, surface coatings to protect parts against wear and corrosion, replacement joints for the human body, cutting tools, and even thermal barriers for the nose-cones of rockets.

GLASS

Glass, that magical material, is also of mineral origin. It is mainly made of silica sand, lime, and sodium or potassium carbonate.

To date the origins of glass, we need first to acknowledge that it exists in three naturally occurring forms that pre-date life on Earth. These are tektite, which is formed by impact from meteorites; fulgurites, which are formed by lightning strikes in sandy soil; and, most common of all, volcanic glass, such as obsidian and basalt. Obsidian was used in the earliest pre-Columbian civilizations.

Glass as a manmade material dates from around 4000 B.C.E., but the industry as such started around the beginning of the Common Era. At this time the Phoenicians were already using blowpipes, and the Romans, instead of importing their glass from Egypt, inaugurated the first glassworks in Rome. Venice subsequently became the European capital of glass, although it lost some of its superiority in the sixteenth century to places like Prague, where the production of intricately engraved Bohemian glass would develop. Nevertheless, it was in Venice, at the turn of the sixteenth century, that clear window glass was first made. The industrialization of glass started in England in the seventeenth century, with the coal-burning furnaces of Robert Mansell, which in turn led to the invention of a much clearer "flint glass" containing lead oxide.

A.M. CASSANDRE, POSTER FOR TRIPLEX GLASS, FRANCE, 1931. ("TRIPLEX GLASS: CRACKS BUT DOES NOT SPLINTER.")

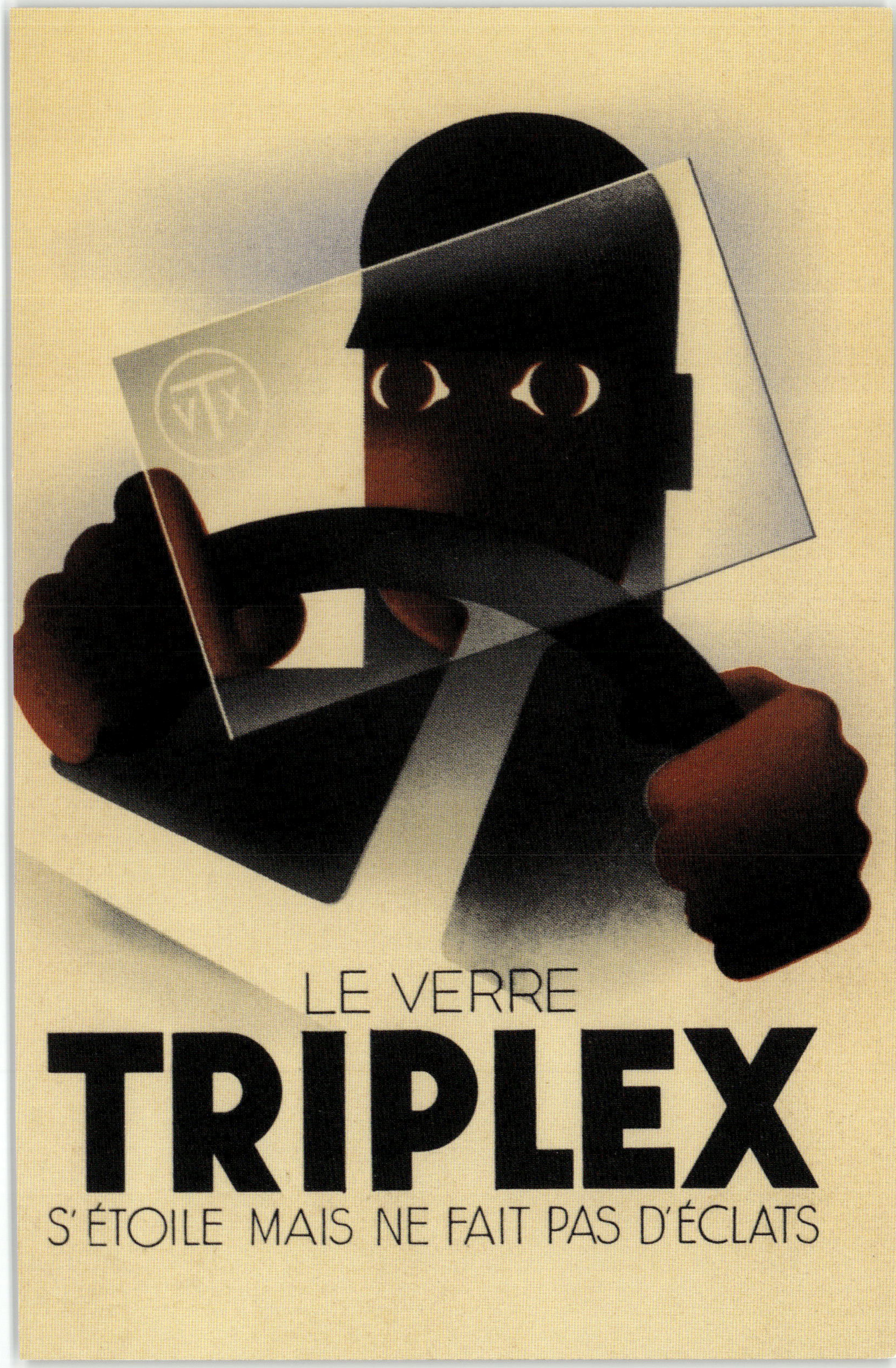
VTX
LE VERRE
TRIPLEX
S'ÉTOILE MAIS NE FAIT PAS D'ÉCLATS

THE MANUFACTURE OF INDUSTRIAL GLASSWARE

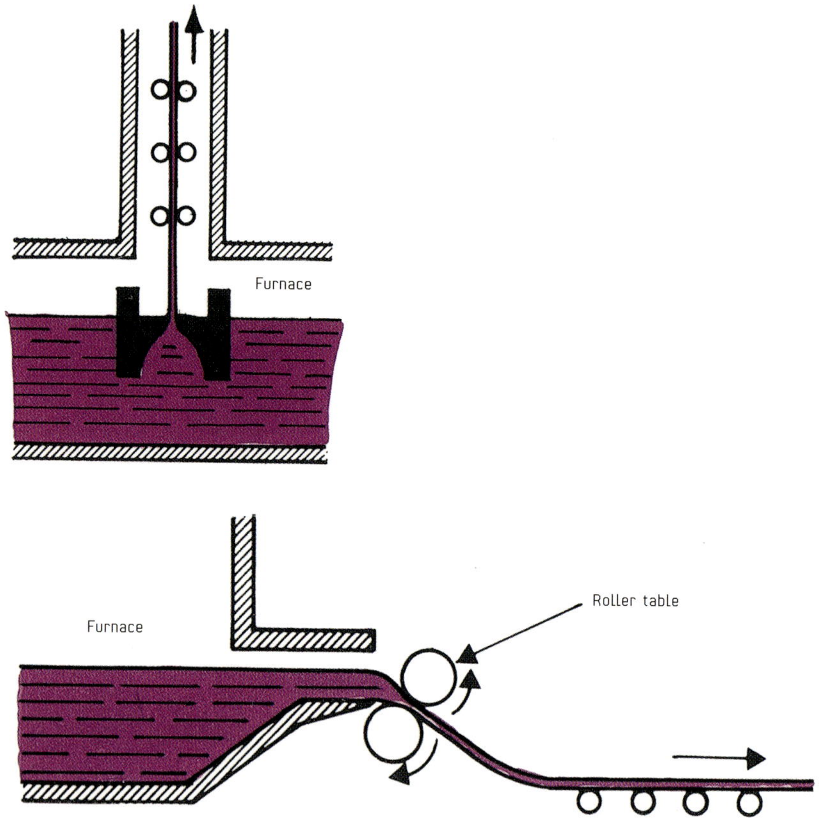

TWO METHODS OF MANUFACTURE OF FLAT GLASS

above, top

VERTICAL DRAWING OF GLASS (THE FOURCAULT METHOD). THE GLASS IS PULLED UP FROM THE MELT BY A SET OF ROLLERS TO THE FLOOR ABOVE, WHERE IT IS CUT INTO SECTIONS.

above, bottom

THE ROLLING PROCESS, SIMILAR TO THAT USED FOR METAL. THE THICKNESS IS DETERMINED BY THE SPACING OF THE ROLLERS.

MOLD PRESSING WITH A SINGLE-PIECE DIE

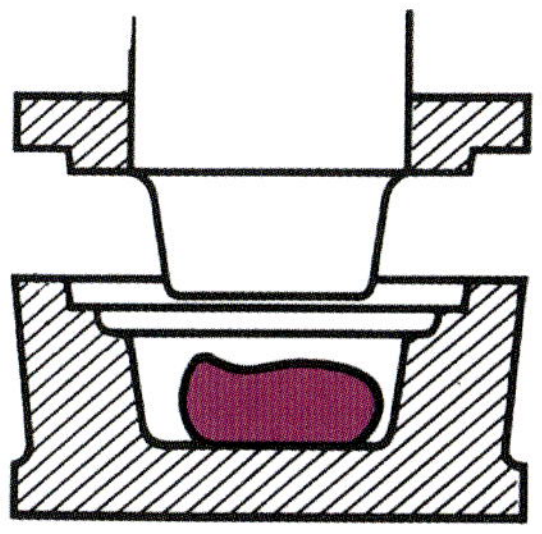

A GOB OF SEMI-MOLTEN GLASS WITH THE SAME WEIGHT AS THE FINISHED PIECE IS DROPPED AT THE BOTTOM OF THE MOLD CAVITY, WHICH IS HEATED TO PREVENT THE GLASS FROM SETTING TOO QUICKLY. THE GLASS WILL NOT ASSUME THE SHAPE OF THE MOLD ON ITS OWN: IT HAS TO BE FORCED TO FILL THE CAVITY COMPLETELY.

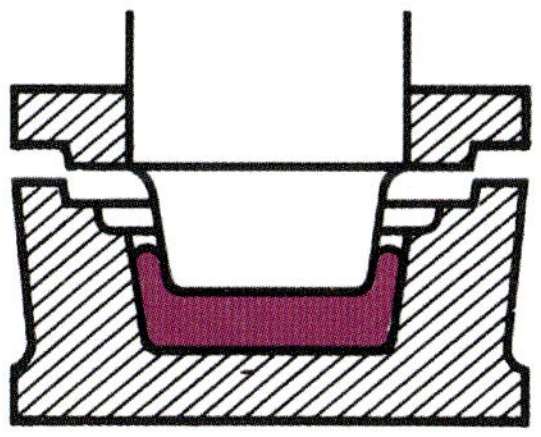

A RING SURROUNDING THE PLUNGER IS LOWERED BEFORE THE PLUNGER ITSELF, SEALING THE LOWER CAVITY AND PREVENTING THE GLASS FROM ESCAPING. THE PLUNGER DESCENDS, FORCING THE GOB OF GLASS INTO THE REMAINING SPACE. AIR ESCAPES THROUGH THE JOINT AT THE TOP OF THE MOLD.

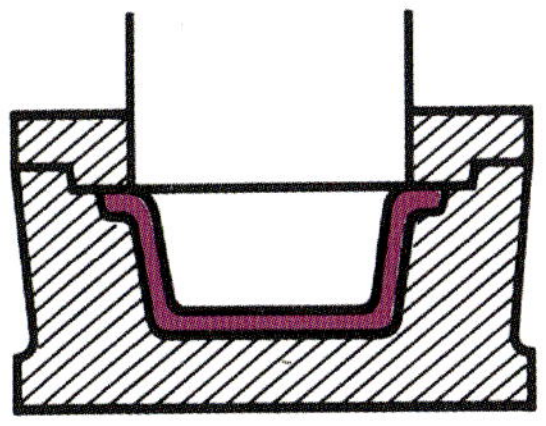

THE PRESS STROKE IS STOPPED BY RESISTANCE FROM THE TRAPPED GLASS. ANY VARIANCE IN WEIGHT OF THE INITIAL GLASS CHARGE RESULTS IN A CORRESPONDING DIFFERENCE IN THE THICKNESS OF THE WALLS OF THE FINISHED PIECE. TOLERANCE CAN BE OF THE ORDER OF ±1MM IN LARGER PIECES.

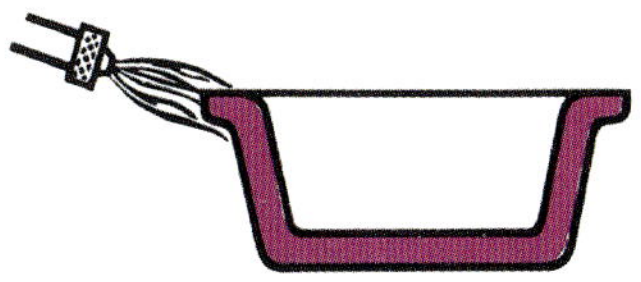

GLASS MAY LEAK THROUGH THE JOINT BETWEEN DIE AND PLUNGER RING, LEAVING A FIN AROUND THE EDGE. ONCE REMOVED FROM THE MOLD, THE SEAM MARK IS DEBURRED WITH A FLAME.

BLOW-MOLDING A HOLLOW GLASS CONTAINER

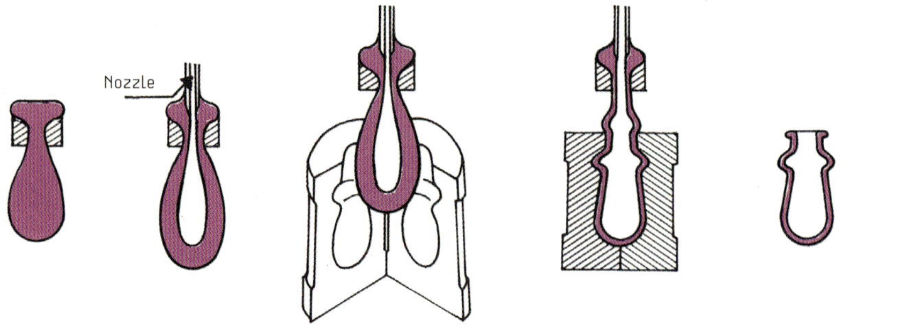

A GOB OF MOLTEN GLASS IS TAKEN FROM THE MELTING FURNACE.

A SMALL AMOUNT OF AIR IS INJECTED.

THE GLASS PARISON IS INTRODUCED INTO THE OPEN MOLD.

COMPRESSED AIR FORCES IT TO TAKE THE SHAPE OF THE MOLD CAVITY.

THE FINISHED PIECE AFTER TRIMMING THE NECK.

MAKING A HOLLOW GLASS CONTAINER BY STRETCHING OR CONSTRICTING A TUBE

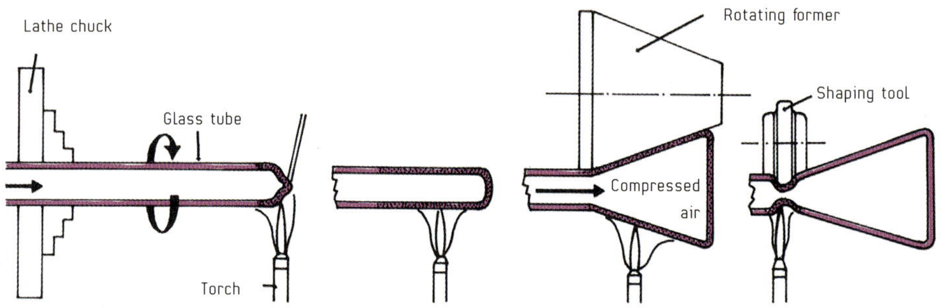

THE TUBE IS MOUNTED ON A LATHE AND THE END IS SEALED TO ALLOW COMPRESSED AIR TO BE INTRODUCED AT THE OTHER END.

THE AREA TO BE FORMED IS HEATED TO SOFTENING TEMPERATURE.

THE ROTATING PARISON IS EXPANDED BY COMPRESSED AIR UNTIL IT MEETS A ROTATING FORMER.

AREAS SUCH AS THE NECK OF THE PIECE CAN ALSO BE SHAPED BY CONSTRICTION.

MAKING A HOLLOW GLASS CONTAINER BY BLOWING THEN PRESSING

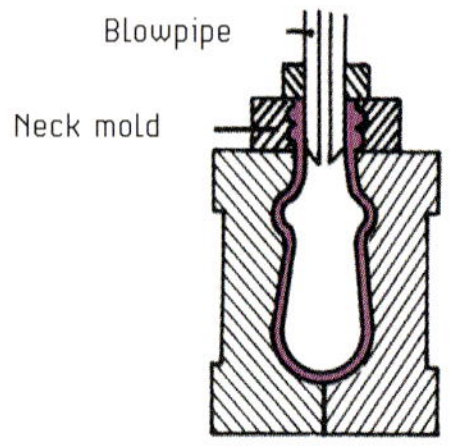

WHERE A SINGLE-PASS MOLDING CANNOT RELIABLY GIVE THE NECESSARY DETAIL (A THREADED NECK FOR EXAMPLE), THE PIECE IS FINISHED BY ROTARY HOT-FORM PRESSING (RIGHT) OR THE APPLICATION OF A NECK MOLD (LEFT).

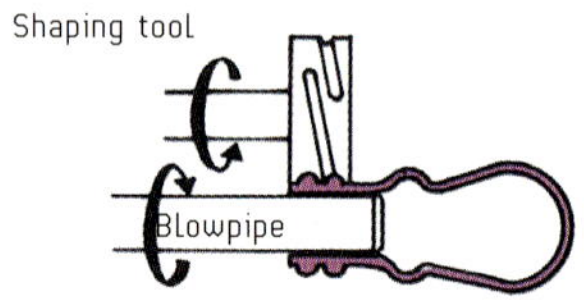

THE PRESS-BLOW PROCESS

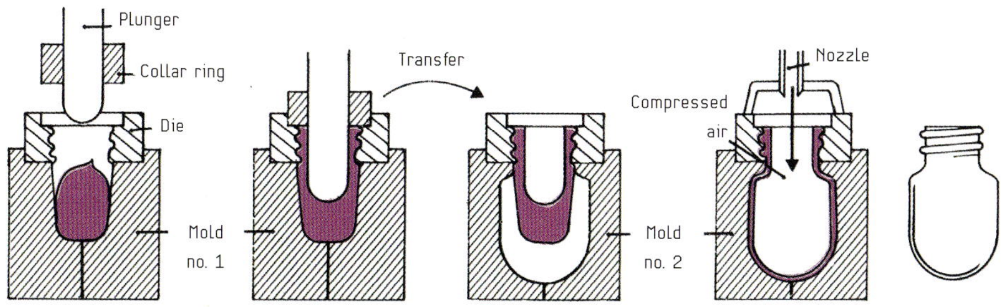

FIRST PHASE: PRESSING STAGE

MOLD NO. 1 IS IN TWO PARTS. THE LOWER CAVITY SHAPES THE GLASS INTO A HOLLOW "BLANK." THE UPPER DIE PRESSES THE TOP OF THE PIECE (THE NECK THREAD IN THIS EXAMPLE), AND IS ALSO USED TO HOLD THE PARISON AND TRANSFER IT TO MOLD NO. 2.

SECOND PHASE: BLOWING STAGE

THE PRESSED SECTION REMAINS CAPTIVE WHILE COMPRESSED AIR IS INJECTED. THE GLASS TAKES ON THE SHAPE AND DIMENSIONS OF THE INNER WALL OF THE LARGER MOLD, GIVING A REASONABLE DISTRIBUTION OF THICKNESS.

FINISHED PIECE

MAKING TELEVISION TUBES BY CENTRIFUGING

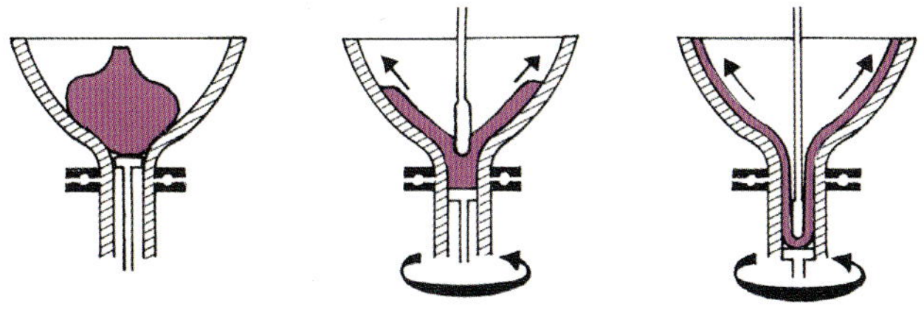

THE GOB OF GLASS IS PLACED AT THE BOTTOM OF THE MOLD.

AS THE MOLD SPINS, A PLUNGER ENCOURAGES THE GLASS TO CREEP UP THE WALLS OF THE MOLD. CENTRIFUGAL FORCE PUSHES THE GLASS AGAINST THE MOLD WALL.

AT THE END OF THE ROTATION, THE GLASS IS DISTRIBUTED EVENLY UP TO THE TOP OF THE CAVITY, UNDER THE COMBINED EFFECT OF THE PLUNGER AND CENTRIFUGAL FORCE.

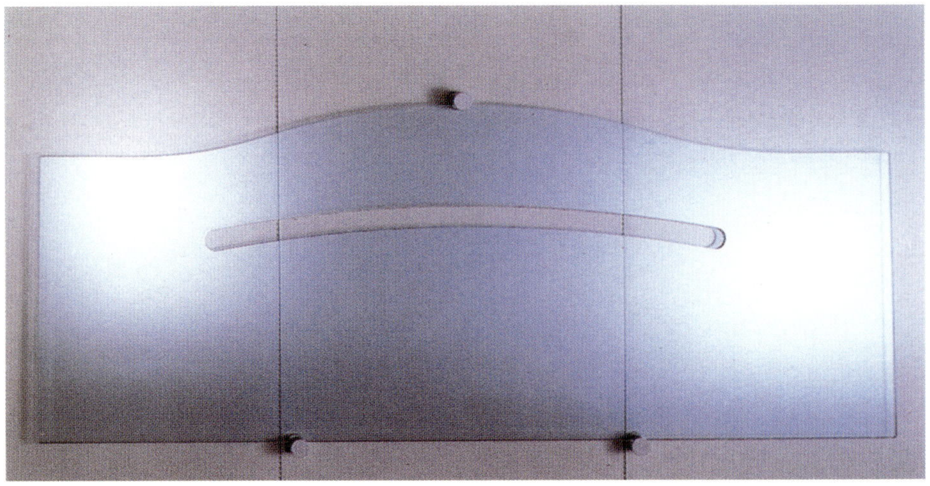

CINI BOERI AND **TOMU KATAYANAGI,** *GHOST* CHAIR, CUT FROM A SINGLE SHEET OF GLASS BY WATER JET THEN SHAPED AND TEMPERED, BY FIAM, ITALY, 1987.

VITRO LUX, "SUB-SURFACE"
ENGRAVING IN SOLID GLASS.
THE DESIGN IS ETCHED ON
THE INSIDE OF THE PIECE WITH
A LASER, POINT BY POINT,
LEAVING THE SURFACE INTACT.
VITRO LASER, GERMANY.

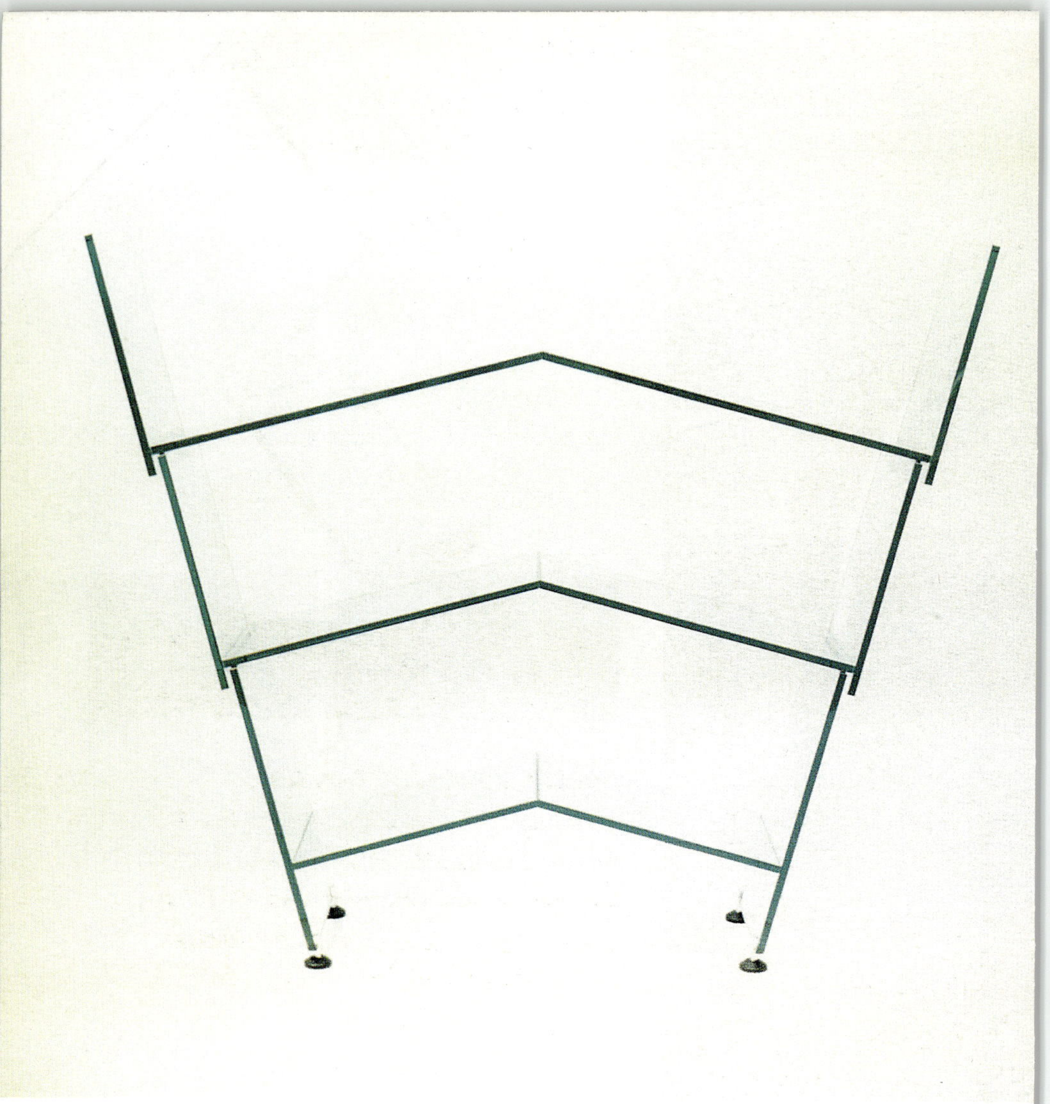

ISAO HOSOE, *ALBERO* BOOKCASE, EDGE-GLUED GLASS, PRODUCED BY TONELLI, ITALY, 1988.

facing page
PHILIPPE STARCK, TABLE FROM THE *LANG* SERIES, PRODUCED BY DRIADE, ITALY, 1987.

Flat glass has been particularly important in the development of industrialized civilization. "Muff" glass was made from the eleventh century by blowing a glass sphere and rolling it into a cylinder, then cutting the ends off, reheating, opening out and flattening the cylinder. "Crown" glass then came in at the beginning of the fourteenth century. This involved blowing a sphere and flattening it, then enlarging the semi-molten disc by spinning it out into a "table."

Meanwhile in France, the mirrors in the Hall of Mirrors at Versailles (1678–83) were made with blown glass, but a few years later in 1687, a revolution in glassmaking was to take place. Bernard Perrot developed a technique for casting glass on a metal table, where it would cool and solidify.

It was in the nineteenth century that important steps were taken in continuous production of flat glass. The basic principal consisted of dipping a metallic "bait" into viscous glass; as the plate is raised, it takes a layer of glass with it, which gradually solidifies. This technique was followed in 1918 by one elaborated by Max Bicheroux, which involved passing semi-molten glass through rollers. A considerable amount of grinding, smoothing, and polishing was needed to make the resulting sheets of glass perfectly transparent. This major disadvantage was eliminated in 1960 with the invention of "float" glass by Pilkington in England. The technique simply involves running molten glass out of the furnace onto a long bath of molten tin. The glass spreads out on the surface of the tin to a completely smooth and parallel "ribbon."

Alongside the technical advances in the production of flat glass over the last few centuries, it is as well to mention the various developments that have made glass today into such an indispensable material in so many areas. Often, this has been a matter of adding strengthening materials.

TAPIO WIRKKALA, SERIES OF GLASSES. THE BUBBLE IN THE BASE IS OBTAINED BY INSERTING A WET ROD INTO THE MOLTEN GLASS DURING MOLDING. PRODUCED BY IITTALA, FINLAND, 1954.

Léon Appert in 1893, for example, addressed the safety issue inherent in using such a fragile material for glazing, by hot rolling a strengthening wire mesh between two layers of glass. In 1903 Edouard Bénédictus invented laminated glass, which was sold under the name Triplex; a layer of cellulose acetate was sandwiched between two sheets of glass.

In 1915, the American researchers Eugene Sullivan and William Taylor of Corning Glass Works invented a type of borosilicate glass that would allow a glass food receptacle to resist the temperature of a household oven; it was sold under the brand name Pyrex.

Since the 1960s, there has been an explosion in the diversity of glassware products, and the working techniques have undergone rapid acceleration. One example is chemical tempering, which, in 1965, replaced the thermal tempering technique used since 1870. The mechanical strength of chemically tempered glass, particularly its resistance to bending forces, is three times higher than with the previous method.

In 1968, the Canadian Norman Franz invented water-jet cutting: a high-pressure jet of water, with abrasive material added, able to cut steel, plywood, textiles, and glass, but perform more gently than a laser. The computer-controlled system was used in the 1980s to cut shapes from flat glass. After heat-molding and chemical tempering, these cut pieces could be used for large-scale production of furniture, such as the Ghost chair designed by Cini Boeri for Fiam in 1987.

Glass in designer products is present all around us, and not just in the countless bottles and containers so intensely sought out by designers. Glass fiber is used in composite materials, textiles, and fiber optics. Photochromatic glass lenses can vary their opacity depending on the intensity of light or electrical current. There are transparent vitro-ceramics, types of glass specially adapted for light sources, and flat screens for computers using sheets of glass only two-hundredths of an inch thick. Soon it will be in our micro-processors.

Quite apart from the technical performance we can expect of glass in future cutting-edge research, it also offers new creative possibilities to designers. The Ghost chair mentioned above is one remarkable example. But the use of cyanoacrylate adhesives, which allow glass to be bonded directly onto glass or other materials, has opened up new vistas for more daring minds. Philippe Starck's Lang table (1987), with aluminum legs stuck onto a glass top, is one such example, as is Ettore Sottsass's superb glassware made at the beginning of the 1990s for the Venini studio in Murano. The Accademia and Albero bookshelves designed by Isao Hosoe and produced by Tonelli in Italy in 1988 show just how far it is possible to go with the simple idea of edge-gluing pieces of glass together.

LITRACON, BRICKS OF LIGHT-TRANSMITTING CONCRETE, A COMBINATION OF OPTICAL FIBERS AND CEMENT. NATURAL LIGHT CAN BE LET INTO A WINDOWLESS ROOM: THE WALLS THEMSELVES BECOME A LIGHT SOURCE. DEVELOPED BY LITRACON, GERMANY.

With such creative possibilities, designers are in their element. Glass has had its master craftsmen just as ceramics has, and over the centuries an impressive list of masterpieces has been produced. The new trends developing at the end of the nineteenth century—Art Nouveau in Belgium and France, the Modern Style in Great Britain and the United States, Jugendstil in Germany and Austria—saw a real revolution in original, small-scale production. This has not yet waned, as proved by the output of a number of master glassmakers today. Many designers, not least Ettore Sottsass, have created some remarkable one-of-a-kind pieces over the last decades. Where large-scale production is concerned, designers like Wilhelm Wagenfeld from Germany or the Finnish designers Kaj Franck, Alvar Aalto, and Tapio Wirkkala have contributed some great classics to the history of design.

METALS

Of the one hundred and sixteen elements in Dmitry Ivanovich Mendeleyev's 1886 periodic table of chemical elements, seventy-five of them are metals. They generally have good mechanical properties, ductility in particular. They are good conductors of heat and electricity.

Copper reached Egypt around the year 3600 B.C.E., as suggested by the discovery of copper tools for cutting wood or leather. The combination of copper ore and tin discovered in the third millennium B.C.E.—presumably by accident—heralded the start of the Bronze Age. In an important step for metallurgy, the casting techniques so familiar to potters working with sand and clay were applied to the new material. Above all, its relatively low melting point allowed successive castings in the same molds carved out of stone, an idea that would eventually lead to the modern mass-production of identical copies from an initial pattern.

As well as the permanent (and therefore re-usable) mold, the casting of bronze gave rise to the technique of sand casting, especially for larger pieces. In this process, the mold is made from a mixture of silica sand and clay, to which bonding and other agents are sometimes added. The cavity into which the metal will be poured is formed by packing sand evenly around a pattern with a "strickle." When casting a bell, a profiled strickle is attached at the central axis and rotated to shape the walls of the mold. A mold can also be built around a pattern made of clay, wood, plaster, or

CITROËN FOUNDRY AT CLICHY,

CASTING MOLDS FOR SMALL

AUTOMOBILE PARTS, FRANCE, 1931.

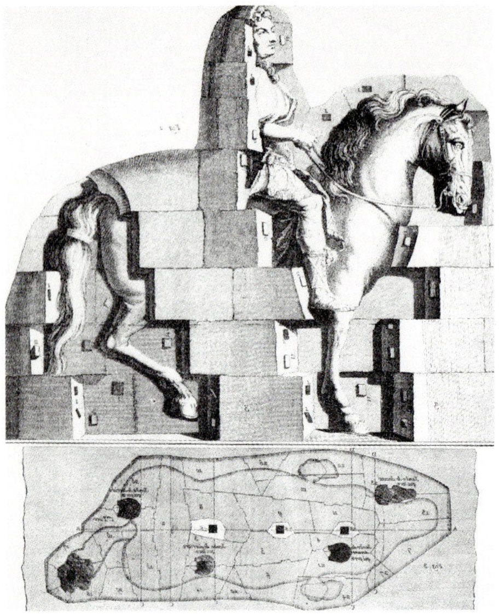

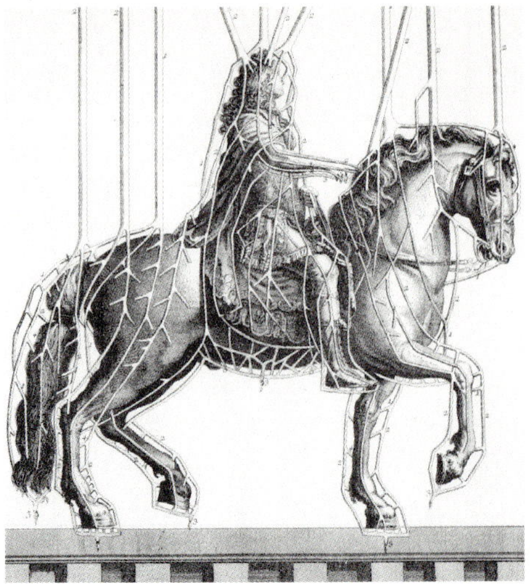

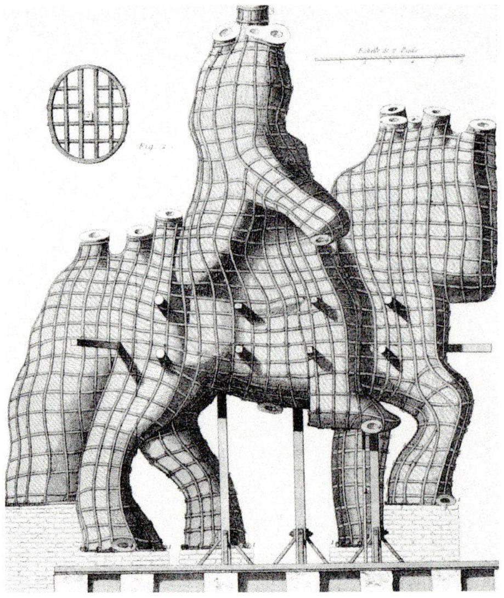

CASTING PLANS FOR FRANÇOIS GIRARDON'S EQUESTRIAN STATUE OF LOUIS XIV ERECTED IN PARIS IN 1699, CAST IN A SINGLE POUR BY JEAN BALTHAZAR KELLER.

left, top
CONSTRUCTION OF PLASTER MOLDS AROUND THE ORIGINAL SCULPTURE. THESE ARE REASSEMBLED TO MAKE A WAX SHELL MODEL, OF THE SAME THICKNESS AS THE FINAL BRONZE.

left, bottom
CONSTRUCTION OF THE INNER CORE WITH ITS SUPPORTING IRON FRAME.

right, top
THE WAX SHELL AROUND A SOLID CORE, LOOKING IDENTICAL TO THE ORIGINAL SCULPTURE. SPRUING IS CARRIED OUT: A NETWORK OF WAX RODS IS ADDED TO PROVIDE FEEDER TUBES, RISERS, AND VENTS FOR THE HOT METAL.

right, bottom
THE WHOLE ASSEMBLY IS COATED IN A REFRACTORY SLURRY, THEN PACKED WITH CLAY AND SAND, AND STRENGTHENED WITH METAL BANDS. THE MOLD IS FIRED, AND THE WAX MELTS AND RUNS OUT. THE MOLD IS READY FOR THE METAL TO BE POURED.

Fig. 3
G
Fig. 2
Fig. 1

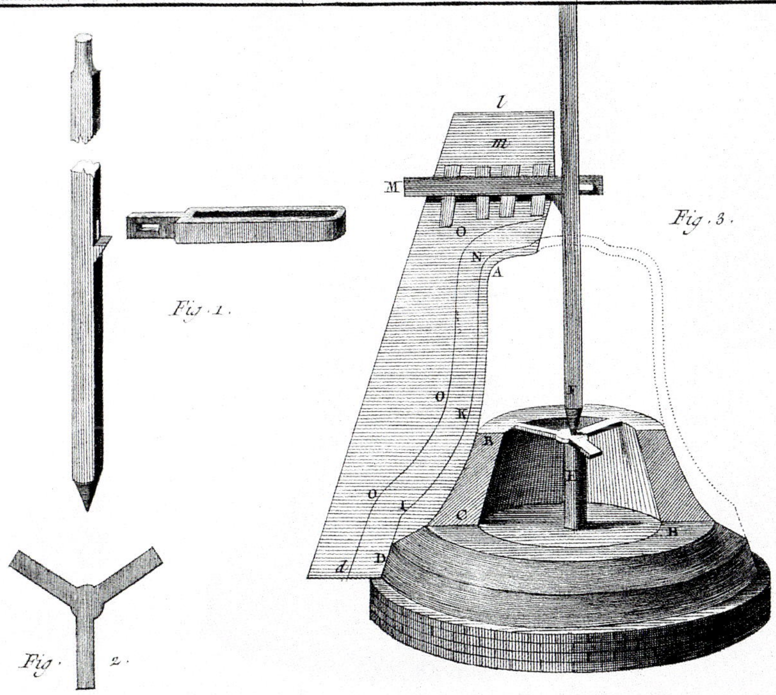
Fig. 1.
Fig. 2.
Fig. 3.
l
m
M
O
N
A
K
B
C
D
d
H

even metal. For a hollow piece, the space inside is taken up with a "core": a complementary shape also made of sand.

An even older technique is that of "lost wax," which was also used for casting bronze. A wax blank is made, representing the final object, and is covered with hemp-reinforced earth. The mold is built up and then heated to melt the wax; the cavity is emptied and molten metal is poured in.

These foundry techniques are applicable to all metals and have progressed remarkably over the centuries. The greatest strides have been made in the iron and steel industry, however, of necessity due to the high melting temperatures involved.

From the beginning of the third millennium B.C.E, or perhaps even earlier, the Sumerians and Egyptians were carving and shaping crystalline iron ore, magnetite (Fe_3O_4) and hematite (Fe_2O_3). No doubt charmed by the metallic sparkle of these materials, they fashioned weights, decorated cylinders, and created kohl sticks that have survived the millennia.

Recently transcribed writings from Al Fayyum, 3500 B.C.E, testify that the Egyptians knew how to extract iron from ore at that time.

IRON AND THE INDUSTRIAL REVOLUTION

The discovery and mastery of iron and steel production and working techniques were a vital part of the Industrial Revolution. Credit is due to the English foundry owner Abraham Darby, whose trials resulted in an enlarged blast furnace in 1709, in which iron ore could be smelted more efficiently by using coke rather than charcoal as fuel.

The now abundant cast iron was remarkably resistant to compression and became widely used as a successor of stone. Everyday objects were produced in increasing numbers, such as stoves and cooking pots, as well as household and municipal furniture and fittings, such as seats, railings, and lampposts, and, of course, machinery. As the nineteenth century approached, iron was used for structural elements in architecture, such as posts and beams. Meanwhile, the technique of puddling was developed by

preceding page

BELL CASTING: SHAPING THE MOLD BY STRICKLING. ENGRAVING FROM DIDEROT AND ALEMBERT'S FRENCH *ENCYCLOPÉDIE*, 1751-1752.

facing page

top

NINETEENTH-CENTURY BLAST FURNACES.

bottom

A ROW OF BLAST FURNACES.

following pages

CASTING MACHINE PARTS IN A FRENCH FOUNDRY, 1950S.

Henry Cort in 1783, which allowed the pig iron obtained from the blast furnace to be turned into wrought iron, by eliminating carbon.

The birth of the steam engine—and with it, the steam hammer—meant that great quantities of iron could be worked, by drop-forging and manual forging, enabling the manufacture of increasing numbers of tools, utensils, and machine parts. Cort also invented a rolling mill with grooved cylinders, which could take blocks of puddle iron and produce constant-width sheets with specific mechanical properties. With smooth cylinders, sheet iron could be produced. The sector of semi-finished materials grew: angle iron, beams of I-shaped or U-shaped section; these were products manufactured with no specific destination, but with standard characteristics. They could be used by architects right from the drawing and calculation stage, making them perfectly suited for work on the gigantic scale now being demanded by industrial civilization. One of the best examples of this new use of is the Eiffel Tower.

By the end of the eighteenth century, a gap was becoming increasingly apparent between the possible applications of cast and wrought iron and the huge expense of purifying iron into steel. Henry Bessemer's converter, patented in 1855, allowed massive amounts of steel to be produced for the first time at a fraction of the cost of previous methods. The mechanical resistance of steel to bending, pulling, compression, and torsion can be varied with the carbon content; it remains very malleable when hot, but can be hardened by quenching in water. It soon replaced puddle iron in a great many applications. Then there came special steels: carbon steel with added manganese or nickel for example, for increased hardness and strength. Meanwhile the methods of working these materials, casting in particular, had to evolve to meet the demands of mass production.

Sand casting, an essentially one-time process, was made more efficient by the use of machines to mold the sand. The pattern, mounted on a plate, is covered by a molding-box filled with sand, and the sand is compacted by a machine-driven piston. The increase in speed for each molding allows a higher output.

TAPPING THE BLAST FURNACE. A TAPHOLE IN THE HEARTH IS STOPPED UP WHILE THE FURNACE IS WORKING, WITH A PLUG MADE OF CLAY AND COKE, THEN UNBLOCKED TO DRAW OFF MOLTEN IRON. ARCELOR, FRANCE

Shell molding (also known as the Croning process, after its inventor), also uses a pattern plate, but the molding sand is coated in thermosetting resin. The metal plate is heated and applied to the molding material, causing the resin to melt and harden into a shell, ready for the metal to be poured.

VIEW OF A CASTING FLOOR AT
THE BASE OF A BLAST FURNACE.
ARCELOR, FRANCE

facing page
top
MOLTEN IRON FLOWING FROM
A BLAST FURNACE.

bottom
ON THE CASTING FLOOR: MOLTEN
IRON ARRIVING FROM THE BLAST
FURNACE.

above

ELECTRIC STEEL MILL. A LADLE CONTAINING MOLTEN STEEL IS TILTED TO FILL INGOT MOLDS OR MAKE A CASTING.

facing page

FILLING AN OXYGEN-BLOWN CONVERTER (CAST IRON TO STEEL) AT THE ACERALIA STEELWORKS, GIJON, SPAIN.

LOADING PHASE OF AN ELECTRIC ARC FURNACE: THE CHARGING BUCKET CAN BE SEEN BEHIND AN INCANDESCENT ELECTRODE, READY TO BE LOWERED INTO THE FURNACE. THE ELECTRIC FURNACE IS MAINLY USED FOR MAKING SPECIAL ALLOYS OF HIGH-QUALITY STEEL. SCRAP IRON AND COMPLEMENTARY METALS ARE MELTED BY AN ARC OF ELECTRICITY THROWN BETWEEN GRAPHITE ELECTRODES.

facing page
REFINING STAGE, ELECTRIC STEEL MILL, LUXEMBOURG.

following pages
left
ROLLING MILL FOR COINING, GIOVANNI BRANCA, *LA MACHINE*, ROME, 1629.

right
ROLLING MILL FOR HEAVY-GAUGE PLATE.

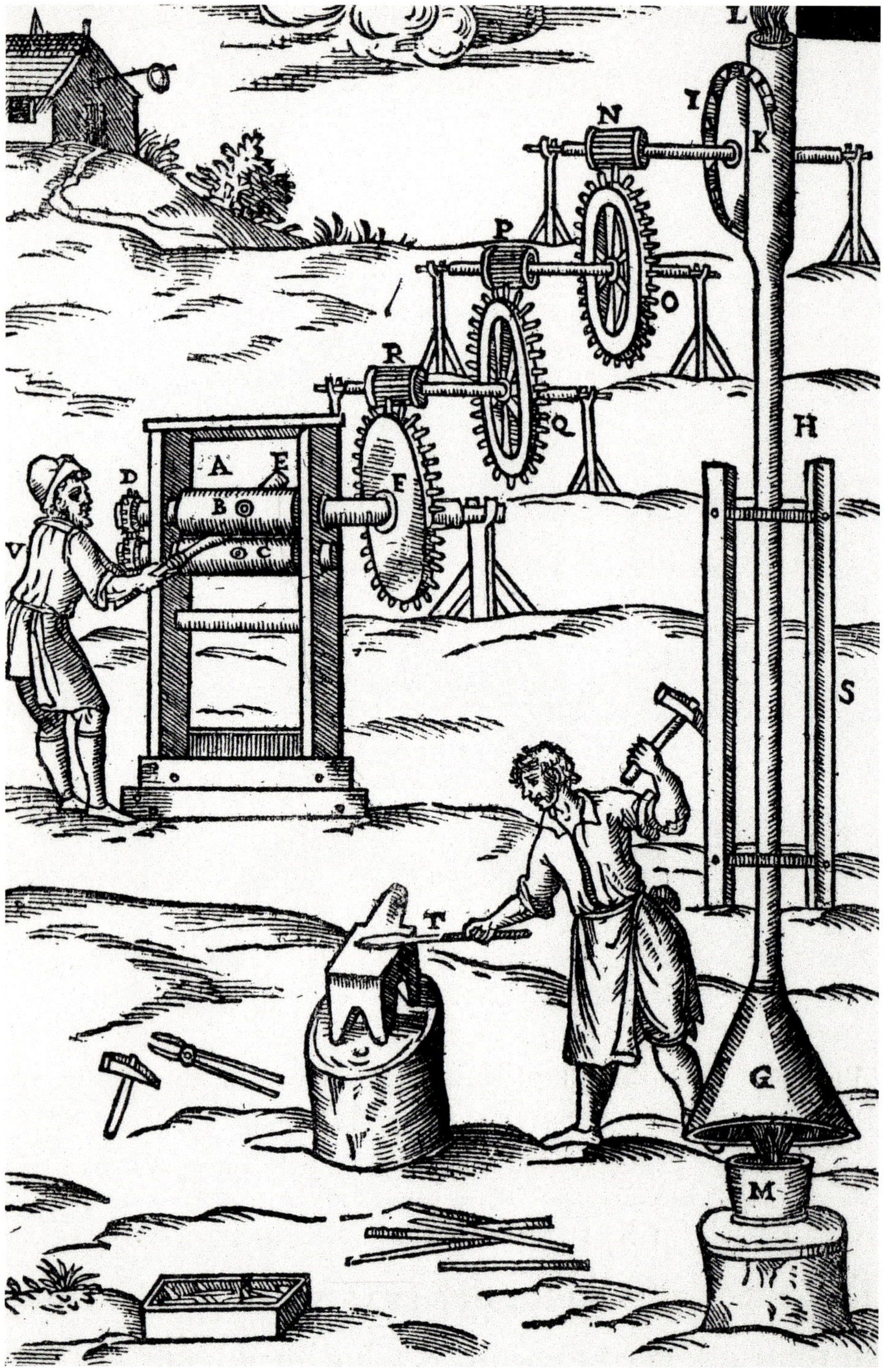
L
N
Y
K
P
O
R
Q
H
A
E
D
B
C
F
V
S
T
G
M

top

COGGING MILL. THE HOT INGOT PASSES THROUGH ROLLERS, REDUCING IT TO SLAB (FOR FLAT PRODUCTS) OR BLOOM (FOR LONG, PROFILED PRODUCTS). FROM THERE, THE STEEL WILL PASS THROUGH A ROUGHING STAND, A REHEATING FURNACE, AND A SERIES OF FURTHER ROLLERS, EMERGING FROM THE FINISHING TRAIN AT THE DESIRED DIMENSIONS.

preceding pages

A WIRE MILL WHERE SEVERAL STRANDS OF BILLET (ITSELF PREVIOUSLY ROLLED DOWN FROM CAST BLOOM) ARE COMBINED. SMALLER-PROFILE PRODUCTS ARE MADE THIS WAY, SUCH AS REINFORCING BARS FOR CONCRETE.

bottom

A LINE OF ROUGHING STANDS IN A CONTINUOUS HOT-ROLLING MILL. SLAB STEEL EMERGES AT THE END AS A SEMI-FINISHED PRODUCT (FLAT OR PROFILED, DEPENDING ON SETUP).

above

A ROLLING STAND.

facing page

CONTROL STATION FOR AN ELECTRIC FURNACE.

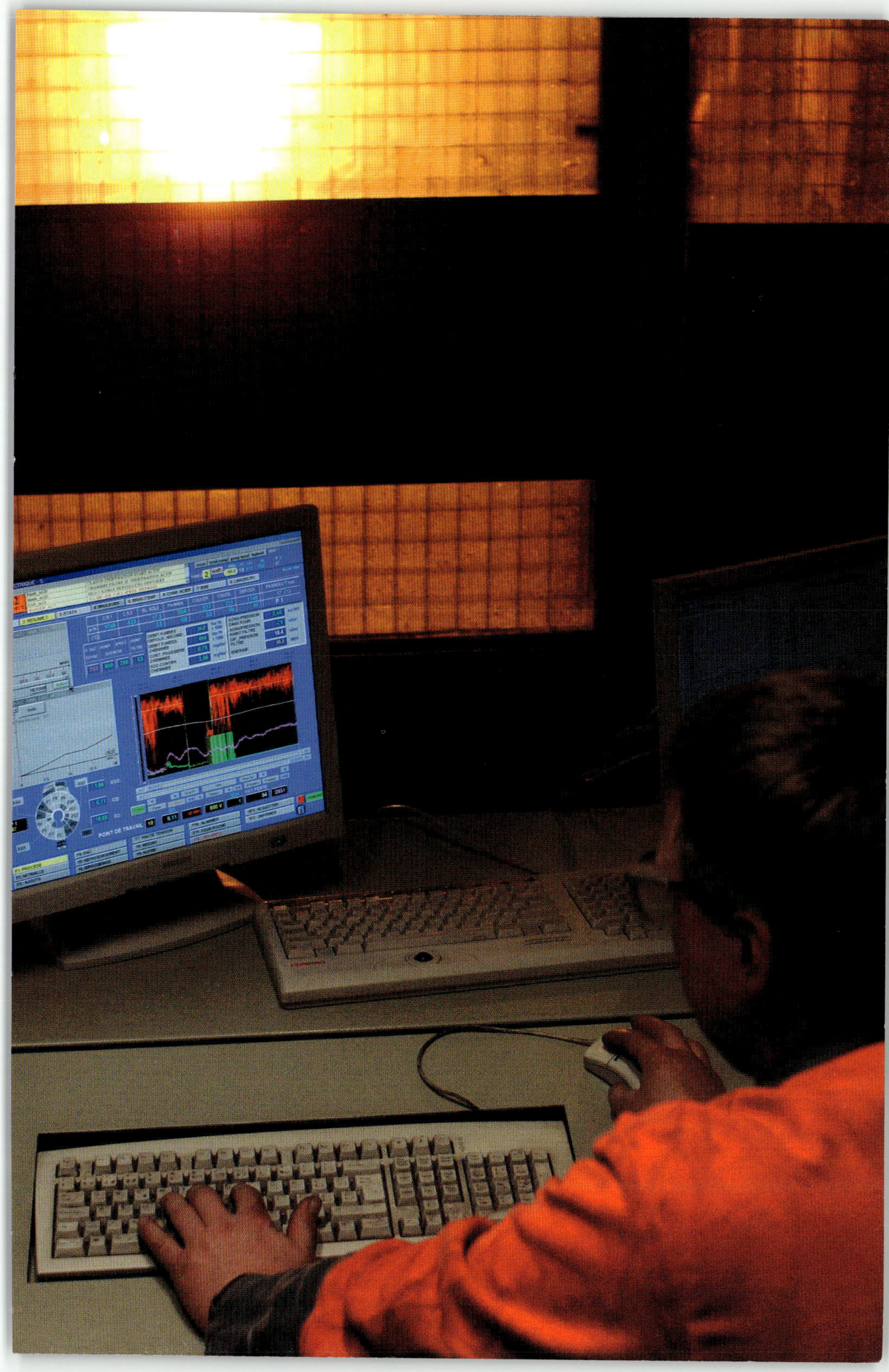

The lost wax method is well adapted to industrial manufacturing, since it can be used to cast smaller pieces with a better surface quality and a high degree of dimensional accuracy. One application is for pieces in extremely hard materials that would be difficult to machine after casting. A ceramic mold, which will harden on firing, can be made around a pattern in wax or another material; the pattern is melted out of the mold or dissolved with solvents. A derivation of the lost wax process is lost foam casting: this involves an expanded polystyrene model that is left in place in the sand and vaporizes as the hot metal is poured. This is used for large pieces, where reusable molds are less of a necessity and more difficult to create.

As casting techniques have developed, so have methods of working metal without stock removal. The drawing or pressing of sheet metal, a derivative of drop-forging, was patented by Jean Palmer in 1848, but was not widely used until the 1930s, when automobile bodywork became aerodynamic. One consequence in the United States was a stylistic influence on domestic appliances, such as streamline vacuum cleaners and even streamline toasters.

With the phenomenal expansion of the automobile industry after World War II, car parts were taken from one machining station to the next by conveyor chains.

Factory pressing usually involves relatively thin sheet metal, especially where car bodywork is concerned. It is now possible to shape pieces with a thickness of nearly four inches (100 mm) by explosion forming: the die cavity is vacuumed and the sheet is blown into shape by detonating an explosive charge under water.

When one considers the apparent simplicity of cold-working tubes, bars, and sheet metal by folding or bending, and the rich variety of shapes that can be obtained, it is easy to see the attraction that these processes had for many furniture designers in the first decades of the twentieth century. The bent steel tubes of bicycle handlebars inspired Marcel Breuer to create his 1925 Club B3 chair, the first of its kind. This inspired a host of designers in the Modern movement to use tubing. Jasper Morrisson's Thinking Man's Chair (1986) shows that the source of inspiration is still as strong as ever.

PRESSING AUTOMOBILE CHASSIS FRAME RAILS USING A 1,400-METRIC TON TOLEDO PRESS, CITROËN FACTORY, ÉPINETTES, FRANCE, 1931.

DIE STAMPING WITH A SINGLE-ACTION PRESS, WITHOUT USING A BLANK-HOLDER.

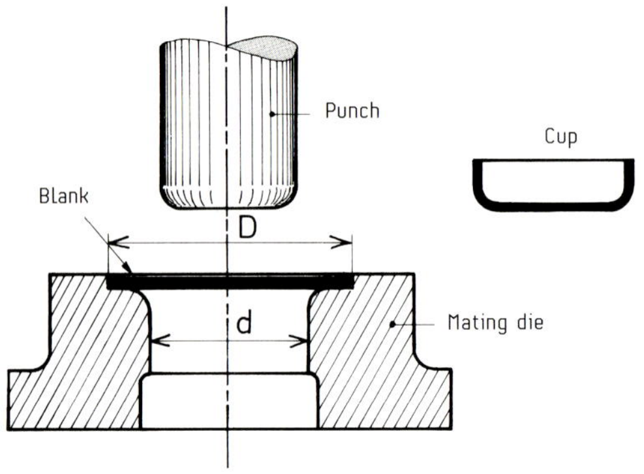

DOUBLE-ACTION STAMPING, USING A BLANK-HOLDER.

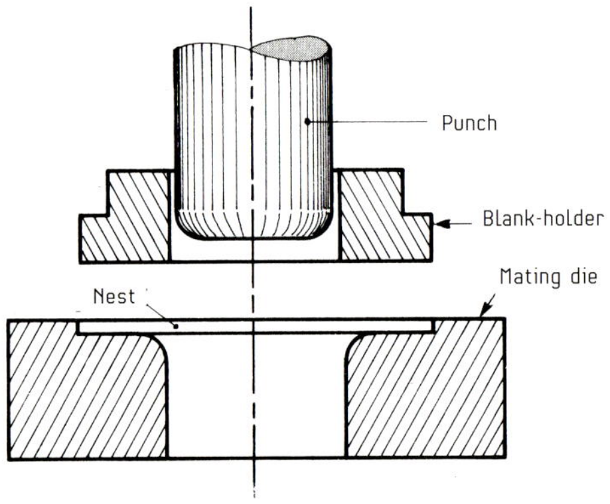

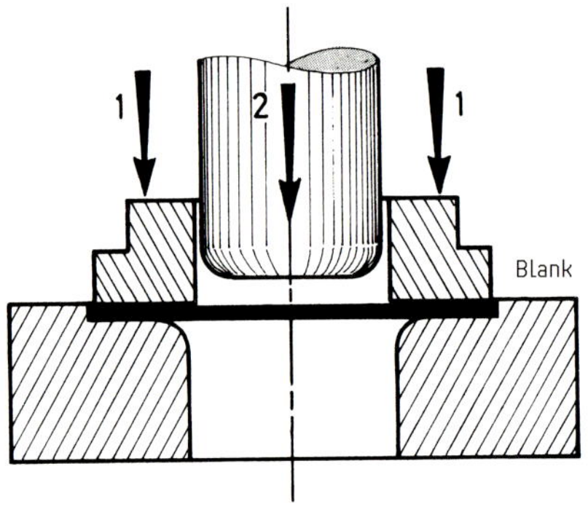

THE BLANK-HOLDER DESCENDS FIRST.

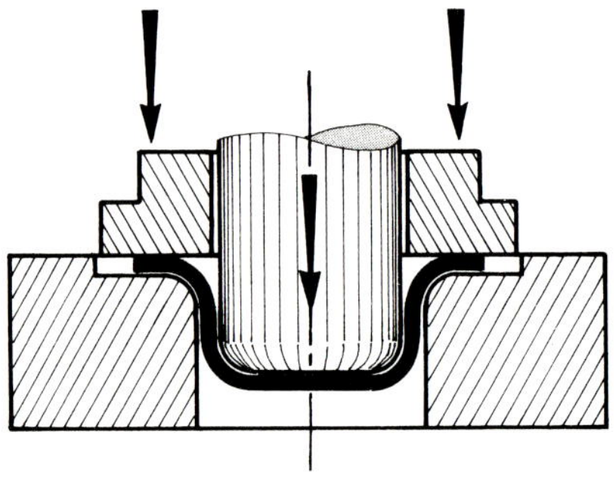

THE BLANK-HOLDER HOLDS THE SHEET IN PLACE WHILE THE PUNCH PRESSES IT.

HPM
HPM
HPM
AULIC PRESS
TURING CO.

preceding page

HYDRAULIC DRAWING PRESS FOR FORMING BASINS, MANUFACTURE DE TOURNUS, FRANCE, 1955.

above and left

FEET ARE WELDED ONTO A STAMPED STAINLESS-STEEL ASHTRAY. WITH PARTIALLY NICKEL-PLATED BRASS VERSION, PRODUCED BY ALESSI, ITALY, AFTER A ONE-OFF DESIGN BY MARIANNE BRANDT AT THE METALLWERKSTATT OF THE BAUHAUS IN WEIMAR, 1924.

above and right

PERFORATION OF THE DECORATIVE FIGURES ON A *KING KONG* METAL BASKET, DESIGNED BY STEFANO GIOVANNONI AND GUIDO VENTURINI. PRODUCED BY ALESSI, ITALY, END OF THE 1980S.

above

STAMPING OF BODY PANELS, RENAULT PLANT, BOULOGNE-BILLANCOURT, FRANCE, 1950S.

facing page

STAMPING SHOP AT A FORD PLANT IN THE 1930S, THE STREAMLINE ERA.

following pages

FAGOR STAMPING LINE AT PSA PEUGEOT CITROËN.MANUFACTURE OF PEUGEOT 1007 AND 206, POISSY, FRANCE, 2004.

ILOT 5

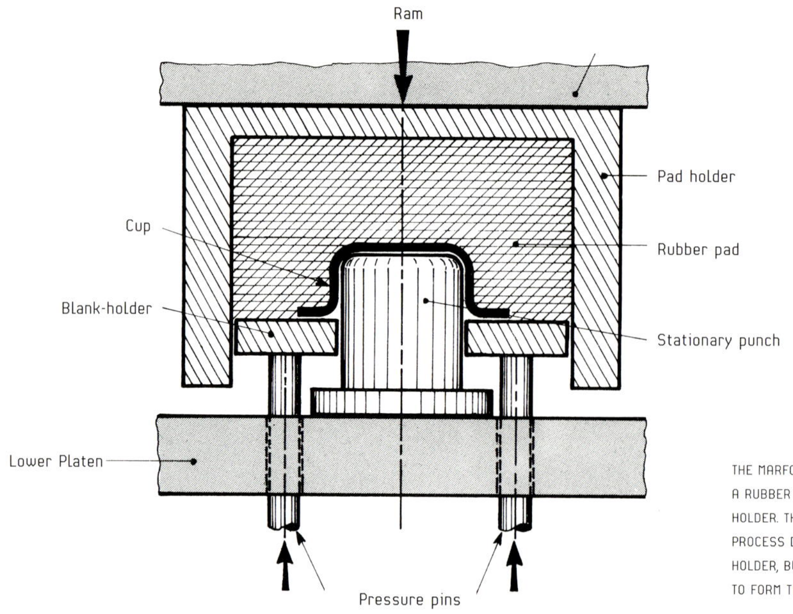

THE MARFORMING PROCESS USES A RUBBER PAD AND A BLANK-HOLDER. THE SIMPLER GUERIN PROCESS DOES NOT USE A BLANK-HOLDER, BUT CAN ONLY BE USED TO FORM THIN, SHALLOW PIECES.

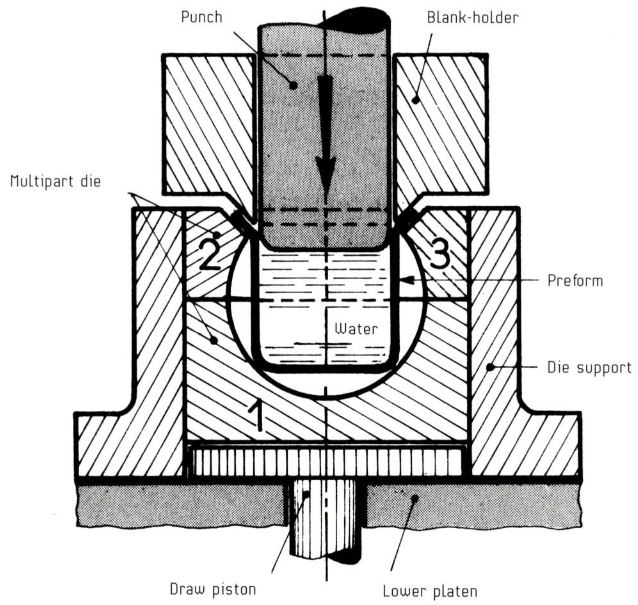

SHAPING SHEET METAL BY HYDROFORMING OR RUBBER PAD HYDROFORMING. THIS METHOD ALLOWS CONCAVE SHAPES TO BE OBTAINED FROM A CYLINDRICAL PREFORM.

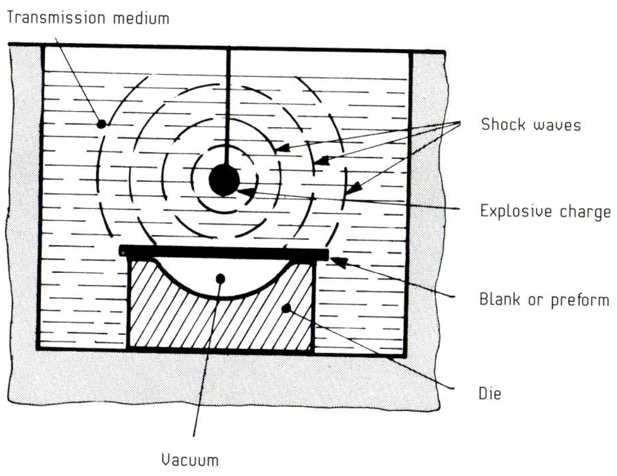

EXPLOSION FORMING. THE MEDIUM FOR TRANSFERRING THE SHOCKWAVE CAN BE LIQUID, POWDER, PASTE OR A GAS; WATER AND SAND ARE MOST OFTEN USED.

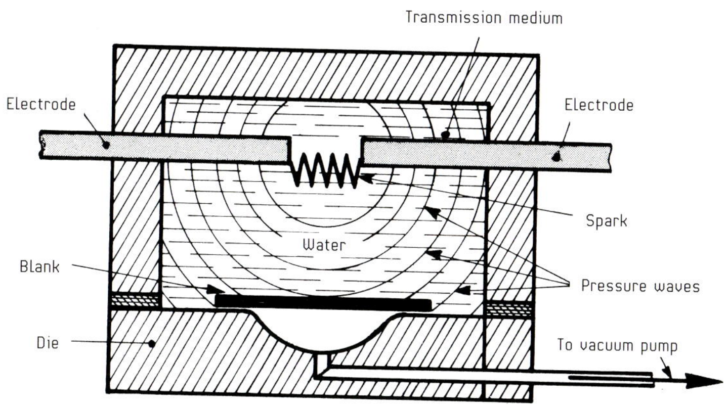

HYDROSPARK FORMING. THE SHOCK WAVE IS GENERATED BY AN ELECTRICAL SPARK PRODUCED BY A CURRENT OF HIGH INTENSITY BUT SHORT DURATION.

Whether in the form of bent sheet metal, as with Jean Prouvé's Chaise Articulée (1924–29), or welded rods, as with Harry Bertoia's Diamond Chair 442 (1952), or the flexible leaves of Ron Arad's Well-Tempered Chair, or expanded metal, as with Shiro Kuramata's How High the Moon (1986)—and there are others—it is steel, semi-fabricated and cold-worked, that has inspired so many designers, and continues to do so today.

Since the Industrial Revolution, the working of iron and steel has undoubtedly been the major influence on shaping techniques of all sorts. These techniques can be used on most metals, and where aluminum and its alloys are concerned, the technical and economic implications are constantly increasing.

ALUMINUM

Given its growing use in all areas, especially in domestic life, it is difficult to imagine that before the discovery in 1754 of naturally occurring alumina, aluminum, the metal that was to be extracted from it was completely unknown. Thanks to its low melting point (1220°F), aluminum and its alloys can be shaped in ways that ferrous metals cannot. They can be cast without using the heat-resistant materials needed for casting iron and steel, which have to be destroyed to remove the finished piece. Polished steel molds receive the liquid metal, either by gravity or under pressure in an injection press. These days the repetitive casting cycle is completely automated. Long pieces of regular section, obtained by extrusion, have become essential basic materials for door and window frames, and domestic living-space extensions.

Again, the list of designers who have found inspiration in the possibilities offered by aluminum is a long one. The suggestive curves of die-cast domestic appliances and utensils in 1930s America were an answer to the aesthetic demands of "streamline."

In the 1960s, die casting was the technique of choice of Charles and Ray Eames for the frames of their famous Aluminum Group chairs. Alberto Meda did the same in 2000 with his Frame chairs, which also make use of extruded and bent aluminum bar. Sand casting can, of course, still be used, when a limited number of pieces of a certain size are required, as

DELPY, POSTER ADVERTISEMENT FOR ALUMINUM, 1930S.

L'ALUMINIUM

above

MOLD FOR GRAVITY-CASTING

ALUMINUM.

facing page

EXTRACTION OF A GRAVITY-CAST

PIECE FROM ITS MOLD.

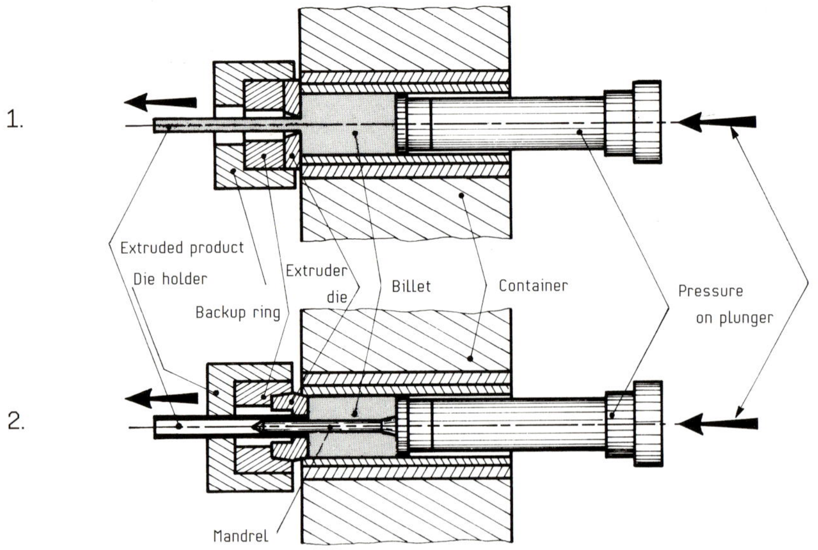

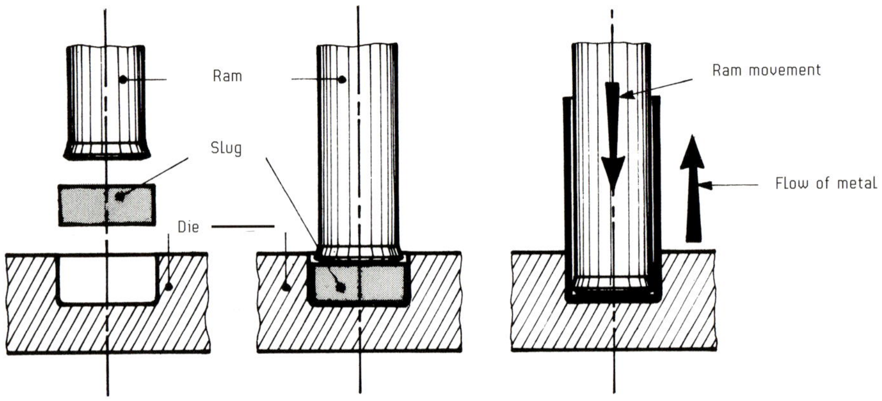

above, top

EXTRUSION.

1. TO OBTAIN FULL BARS.

2. TO OBTAIN TUBES.

above, bottom

IMPACT EXTRUSION. PRESSURE IS EXERTED ON THE WHOLE SURFACE OF THE RAM.

facing page

JEAN PROUVÉ, SECTION OF AN ALUMINUM EXTRUSION FOR STRENGTHENING AND SUPPORTING PLATE GLASS, CIMT (COMPAGNIE INDUSTRIELLE DE MATÉRIEL ET DE TRANSPORT), FRANCE, 1957-1958.

following pages

RON ARAD, *BLO VOID III*, POLISHED ALUMINUM, 2005.

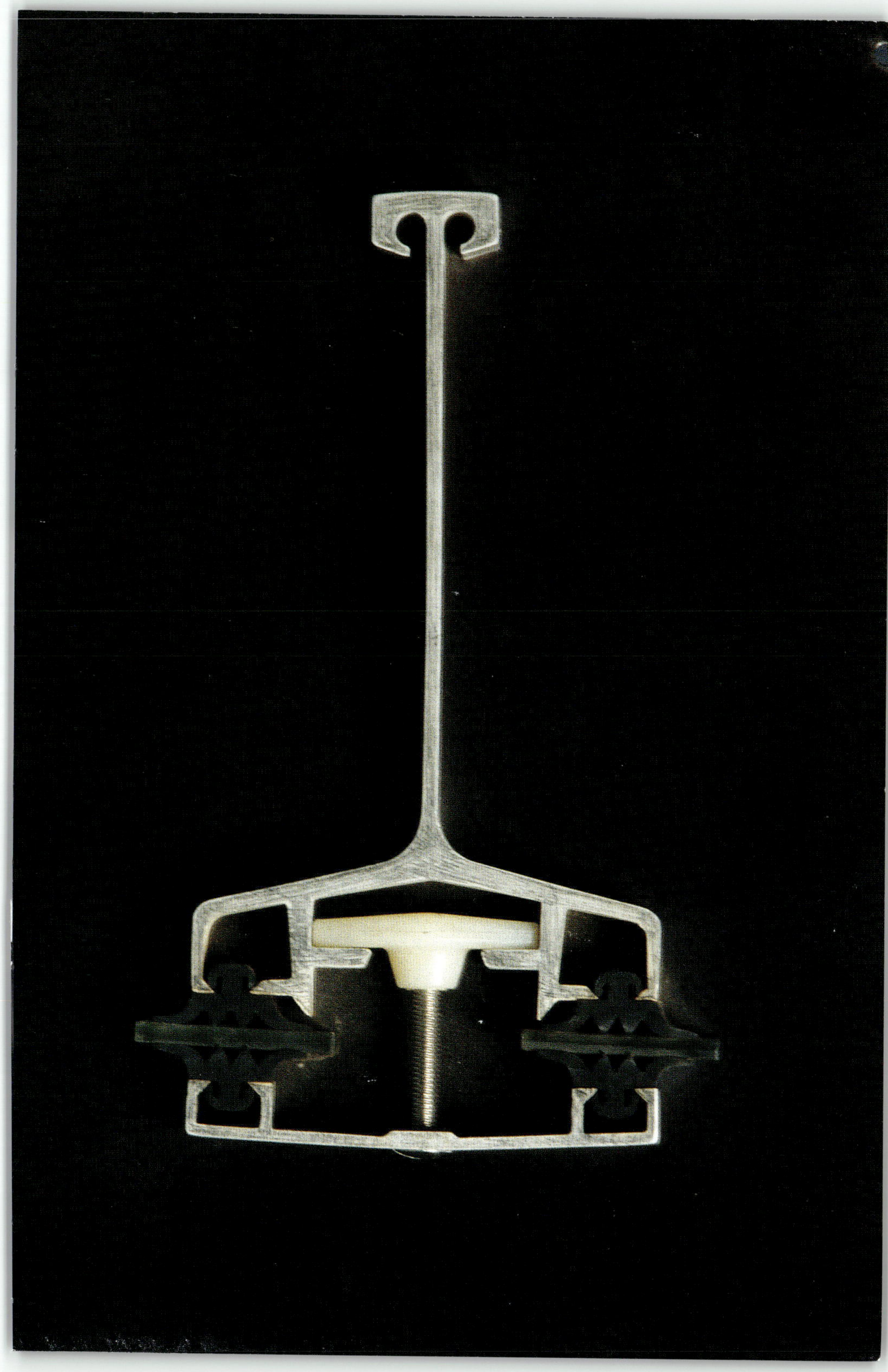

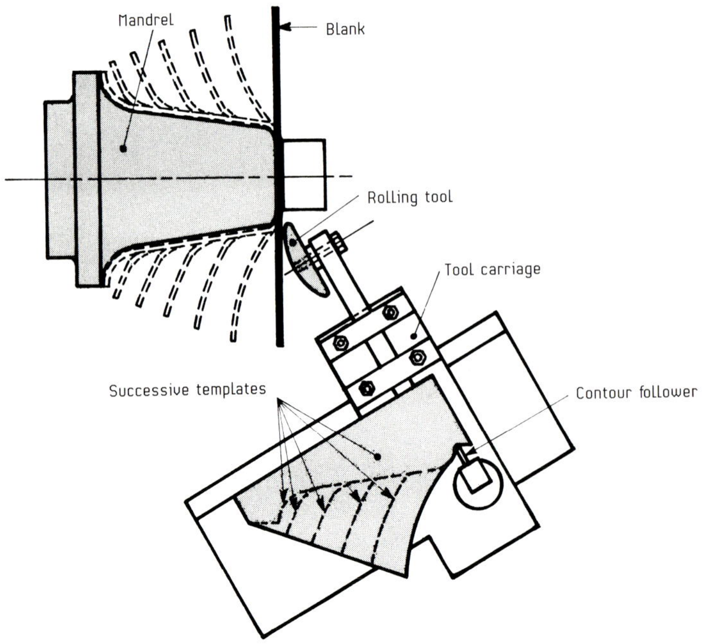

FORMING SHEET METAL
BY MECHANICAL
SPINNING.

ROGER TALLON,

"LES CRYPTOGRAMMES",

LATHE-SPUN ALUMINUM SEATS AND TABLE, 1969, FOR MOBILIER NATIONAL, PARIS. A PLASTIC VERSION WAS PRODUCED IN 1973 FOR LIP, FOR WHOM TALLON HAD DESIGNED A SERIES OF WATCHES AND TIMEPIECES.

right

SPINNING A CIRCULAR BLANK OF LIGHT SHEET METAL (ALUMINUM FOR EXAMPLE) TO OBTAIN A HOLLOW SHAPE.

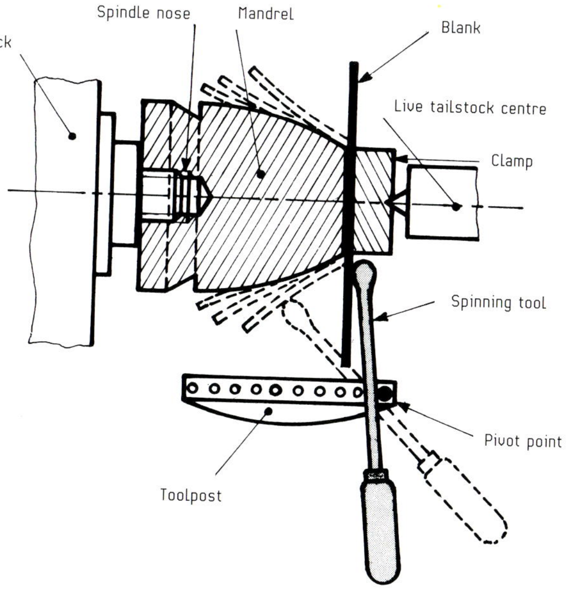

ROGER TALLON, SPIRAL STAIRCASE AND "MODULE 400" FURNITURE, PRODUCED BY LACLOCHE, PARIS, 1966.

was the case with Roger Tallon's Module 400 furniture and spiral staircase, made in 1966 for Galerie Lacloche. Pressed aluminum sheet found one of its most famous applications in 1938 with the Landi chair designed by Hans Coray.

THE UNCERTAINTIES OF RESEARCH

Metallurgy sometimes has surprises in store, even with supposedly well-known materials. But if we don't always have a ready answer or a workaround for a problem, at least today's physicists are well prepared to find one. Sometimes the question arises of whether to replace a metal with a "filled" polymer or a "composite," which can be given the necessary characteristics by combining, for example, mechanical strength with resistance to thermal shock. These attempts do not always come together; sometimes they fail. In early 1971, following a long period of research, hopes of substituting another material in jet engines for aircraft came to nothing.

Research into the extraction and uses of metals of ultra-high purity was initiated in France by Professor Georges Chaudron in the 1950s, and continues to this day. When traces of other elements are removed from magnesium, aluminum or iron, or reduced to an extremely low level (one part per billion), using a combination of purification techniques, the behavior of these metals can change radically.

Metals for which we now have little use will probably find applications in the future that today we can scarcely imagine. Germanium, for example, was just a curiosity until the 1950s, when it was widely used as a semiconductor. Today it has been largely replaced in transistors by silicon.

Likewise, titanium was discovered in England in 1791 by William Gregor, but an application was only found for it recently. It is currently used in aeronautics, heat exchangers in power stations, and medical prosthetics. Titanium alloys are resistant enough to be used for hip replacements, while their lightness and incredible mechanical strength—particularly their resistance to bending—make them ideal for many other applications, one common example being eyeglass frames.

SHAPE MEMORY MATERIALS

"Smart materials" possess intrinsic functions that enable them to behave either as a sensor or as an actuator. They can spontaneously change their physical properties—shape, conductivity, color, viscoelasticity, etc.—in response to a natural or manual trigger, such as a change in temperature, the presence of a magnetic or electrical field, or the application of stress. In short, the material responds to an external stimulus. Actuators respond with a mechanical action or a change in shape, color, or transparency, indicating a change in the environment and prompting corrective action.

There are three categories of smart materials:

- Piezoelectric materials generate an electric voltage when they are stressed, and vice versa. The best-known piezoelectric material is the quartz crystal used in clockmaking.
- Magnetostrictive materials undergo a change of physical dimensions when subjected to a magnetic field; they can be used either as actuators or as sensors.
- Shape memory alloys are "programmable" metal alloys, which can assume a different shape under certain physical stimuli, such as a change in temperature, and return to their initial shape when the stimulus is removed.

The first observation of the shape memory effect was by American researchers L.C. Chang and T.A. Read in 1932, while studying gold-cadmium alloys. The same effect was observed in 1938 with a copper-zinc alloy, but it was not until William J. Buehler's work on a nickel-titanium alloy, Nitinol, in 1963, that the phenomenon was more seriously researched. The first industrial uses were quick to follow: Raychem Corporation began work on an alloy called Tinel in 1967, which was used to make hydraulic couplings in the F-14 Tomcat fighter plane. These materials have been used in a variety of industrial domains since then. From aeronautics to biomedical applications, and from the automobile industry to clothing, they fulfill a range of functions.

Here are some examples of their use:

- Orthodontic arch wires
- Bone staples and strengthening bone implants
- Self-expanding coronary stents. These are fed along an artery via a catheter, to a constricted or weakened point, and expand with the heat of the body.
- Surgical instruments, especially guide-wires for endoscopy.

The food and paramedical sectors have also found uses for shape memory alloys. A tiny device invented by the French company Imago, called a *thermomarqueur*, detects whether there has been an interruption in the cold chain, i.e., a fluctuation in temperature outside the safe range for a given product. A spring made of shape memory alloy expands at a given temperature, pushing a little red ball into view.

The use of shape memory alloys (SMAs) is constantly expanding in the aeronautical and aerospace industries.

The solar panels of some satellites use one-way SMA components. Once in space, the panel opens under the action of a heat source. Sojourner, the rover robot that was sent to Mars, had a two-way electrical actuator that used Nitinol wire to rotate the cover glass on its solar cells.

Some more prosaic applications have been found for SMAs in everyday products. One of the best known is in underwires for bras. The reversible deformation (superelasticity) of the alloys makes the fit more comfortable and means the underwires don't bend in the washing machine.

SMAs have also been very successfully used in spectacle frames. Frames incorporating SMAs have been produced commercially since 1996. The superelasticity of NiTi alloys, or copper alloys (which are more flexible), allows the frames to undergo quite considerable deformation and still return to their original shapes. These examples show just how important shape memory alloys are becoming in industrial design. The same is true of architecture, where SMAs have gained a strong foothold and have an essential role to play. One major application in this sector is to use the damping properties of SMAs for protection against earthquakes.

A number of artists have shown interest in the new possibilities offered by shape memory alloys. Since the advent of avant-garde art in the early twentieth century, the artist's studio has at times been a laboratory of the future: a place where new materials and techniques can be experimented with, free from the concerns of economy and productivity inherent in the industrial world. An artistic interest in SMAs can be enriched by active scientific research: witness the partnership in Switzerland between the sculptor Etienne Krähenbühl and the physicist Rolf Gotthardt. Their work with SMAs includes Onibaba, a reed plantation in nickel-titanium wire, which opens up as you approach.

PLASTICS

The history of plastics can be charted by starting with the discovery of vulcanized rubber. In 1840, Charles Goodyear discovered (perhaps accidentally) that adding sulfur to natural rubber at high temperature would increase its strength yet preserve its elasticity. The same mixture with a very high proportion of sulfur gives ebonite, a hard material that can be molded. All the synthetic materials that followed in the nineteenth century were the result of chemical treatment of natural resins.

- Celluloid (nitrocellulose plasticized with camphor) was patented in 1870 by two brothers, Isaiah and John Wesley Hyatt, in response to a competition launched in 1863 to find a substitute for ivory in the manufacture of billiard balls. The high flammability of celluloid is largely responsible for its gradual disappearance as a marketable product; one of the only celluloid items still commonly available today is the table tennis ball.
- Artificial silk, developed by Hilaire de Chardonnet between 1878 and 1890, used cellulose obtained from cotton or wood; the resulting nitrocellulose filaments were drawn through spinnerets, dried, and twisted.
  In England in 1891, Charles Cross, Edward Bevan, and Clayton Beadle treated cellulose with carbon disulfide and dissolved it in caustic

above and following pages

LAURENT MASSALOUX,

NIGHTLIGHT INCORPORATING
A SHAPE-MEMORY DEVICE. WHEN
THE BULB IS LIT, THE HEAT CAUSES
THE COVER TO RISE.

soda, and obtained viscose. This could be spun, and was the forerunner of both rayon and fibranne, which dominated the textiles market until the end of World War II. In 1898, the Englishman C.H. Stearn patented a transparent film obtained from viscose, but it was not until 1912 that the Swiss textiles engineer Jacques Brandenberger launched Cellophane as a trademarked product.

- Casein plastics were discovered in Germany in 1897 by Adolph Spitteler and Wilhelm Krische. The basic process, of condensing and hardening casein (the protein found in milk), was repeated and marketed under various names in Europe and the United States, including Galalith, Erinoid, Aladdinite, Karolith, and Kyloid. The resulting material was white, but was easily colored. It was used largely for making buttons and was a serious competitor of celluloid when it came to replacing natural tortoiseshell, horn, or ivory.

In America in 1909, the Belgian immigrant Leo Hendrik Baekeland, after seven years of research, created the first completely synthetic material: Bakelite. The molecules—phenol and formaldehyde—would never have been combined under natural circumstances: this material could only have been made in a laboratory. It marked the beginning of a great adventure.

How does the growing family of man-made materials stand today? Firstly, they all have one thing in common: their origin is indirectly organic, since they use carbon compounds. They have giant molecules, macromolecules similar to those that make up living tissue.

They are of three physico-chemical types: thermoplastics (TP), thermosets (TS), and mixtures of the two (TP/TS). The difference between them is due to the way the monomers are linked to form a macromolecule.

Macromolecules can be composed of anything between a few hundred and several hundred thousand monomers. These monomers may be identical (homopolymerization) or of more than one sort (copolymerization). Also, the monomers may join by polyaddition, that is, without leaving a residue, or by polycondensation, in which case they form a byproduct (usually water). In the case of thermoplastics, the polymers take the form of a chain, with monomers linked end-to-end like train carriages. The macromolecules are held together by electrostatic forces that, as with a magnet, are weakened by the application of heat. The macromolecules are liberated, and the material liquefies. When the material cools to its initial temperature, it re-solidifies.

With thermoset polymers, the macromolecules are bound together chemically in a three-dimensional lattice, forming one enormous molecule. The reaction is irreversible, and the substance is infusible (when heated, it will eventually disintegrate, but will not melt).

The raw materials used in the manufacture of synthetics are natural, and therefore high in carbon: cellulose, coal, and particularly natural gas and petroleum. Silicones are a special case, since they are a combination of carbon compounds and silicon compounds.

Synthetic materials can be organized by family, as in the table on the following pages:

ABBREV.	TYPE	CHEMICAL FAMILY	CHEMICAL NAME	DENSITY	MELTING POINT °F
ABS	TP	Acrylics/Styrenics	Acrylonitrile Butadiene Styrene Copolymer	1.10	455
AMMA	TP	Acrylics	Acrylonitrile Methyl Methacrylate	1.14	500
CA	TP	Cellulosic	Cellulose Acetate	1.30	266
CAB	TP	Cellulosic	Cellulose Acetate Butyrate	1.26	284
CN	TP	Cellulosic	Cellulose Nitrate	1.28	262
CP	TP	Cellulosic	Cellulose Proprionate	1.30	277
EC	TP	Cellulosic	Ethylcellulose	1.22	306
EP	TS	Epoxides	Polyepoxides	1.40	—
EPT	TP	Polyolefins	Ethylene Propylene Terpolymer	1.02	428
ETFE	TP	Fluorinates	Ethylene-tetrafluoroethylene Copolymer	1.48	536
EVA	TP	Polyolefines	Ethylene Vinyl Acetate Copolymer	1.05	446
FEP	TP	Fluorinates	Fluorinated Ethylene Propylene	1.32	482
MF	TS	Aminoplastics	Melamine Formaldehyde	1.50	—
PA	TP	Polyamides	Polyamide	1.15	500
PAN	TP	Acrylics	Polyacrylonitrile	1.08	482
PC	TP	Polycarbonates	Polycarbonate	1.20	509
PBTP	TS/TP	Polyesters	Polybutylene Terephthalate	1.32	536
PCTFE	TP	Fluorinates	Polychlorotrifluoroethylene	1.98	603
PE	TP	Polyolefines	Polyethylene	0.94	428
PES	TP	Sulfonics	Polyethersulfone	1.34	644
PET	TP	Polyolefines	Polyethylene Terephthalate	1.37	491
PETP	TS/TP	Polyesters	Ethyl Polyterephthalate	1.34	554
PF	TS	Phenolics	Pheno Formaldehyde	1.45	—
PIB	TS	Polyolefines	Polyisobutylene	1.08	464
PMMA	TP	Acrylics	Polymethylmethacrylate	1.20	500
PMP	TP	Polyolefines	Polymethylpentene	1.04	536
PI	TS	Polyimides	Polyimide	1.60	—
POM	TP	Acetals	Polyoxymethylene	1.42	464
PP	TP	Polyolefines	Polypropylene	0.92	428
PPO	TP	Polyphenylene Oxides	Polyphenylene Oxide	1.30	518

ABBREV.	TYPE	CHEMICAL FAMILY	CHEMICAL NAME	DENSITY	MELTING POINT °F
PPS	TP	Polyphenylene sulfides	Polyphenylene Sulfide	1.35	581
PPSU	TP	Sulfonics	Polyphenylene Sulfone	1.50	536
PS	TP	Styrenics	Polystyrene	1.05	464
PSU	TP	Sulfonics	Polysulfone	1.24	626
PTFE	TP	Fluorinates	Polytetrafluoroethylene	2.10	608
PU	TS/TP	Polyurethanes	Polyurethane	1.26	428
PVAC	TP	Vinyls	Polyvinyl Acetate	1.30	410
PVAL	TP	Vinyls	Polyvinyl Alcohol	1.28	302
PVB	TP	Vinyls	Polyvinyl Butyral	1.30	320
PVC	TP	Vinyls	Polyvinyl Chloride	1.35	266
PVC/ABS	TP	Vinyls	Mixture of Polyvinyl Chloride/ABS	1.22	320
PVC/AC	TP	Vinyls	Vinyl Chloride/Vinyl Acetate Copolymer	1.32	248
PVC-C	TP	Vinyls	Post-chlorinated PVC	1.34	248
PVC/E	TP	Vinyls	Mixture of Polyvinyl Chloride /Chlorinated Ethylene	1.21	306
PVDC	TP	Vinyls	Polyvinylidene Chloride	1.38	306
PVDF	TP	Fluorinates	Polyvinylidene Fluoride	1.42	363
PVFM	TP	Vinyls	Polyvinyl Formal	1.48	338
SAN	TP	Styrenics	Styrene acrylonitrile copolymer	1.08	446
SB	TP	Styrenics	Styrene Butadiene Copolymer	1.04	464
SI	TS	Silicones	Silicone	1.10	—
UF	TS	Aminoplastics	Urea Formaldehyde	1.48	—
UP	TS/TP	Polyesters	Unsaturated Polyester	1.40	455
UC/P	TP	Vinyls	Polyvinyl Chloride/Polypropylene Copolymer	1.20	320

Like the natural resins improved in the nineteenth century, a large number of the twentieth century's completely synthetic materials have interesting histories. In the following two examples examined below, the war effort—always a major catalyst for invention—played a part.

The story of synthetic rubber began in England in 1860, when Charles Hanson Greville Williams distilled natural rubber and determined that isoprene was one of its monomers. Hofman, an engineer with the German company Bayer, polymerized dimethylbutadiene in 1916 to make methyl rubber (which turned out not to be resistant to oil, and therefore not ideal for making tires). A series of improvements led to the production of Buna rubber by IG Farben in 1929. By the start of World War II, Buna rubber was being produced at a rate of 175,000 metric ton per year. Japan's occupation of Southeast Asia during the war denied the United States access to the bulk of the world's supply of natural rubber; the US, therefore, had to perfect a substitute quickly. The result was styrene butadiene rubber, SBR, which is still the most widely used synthetic rubber in the world.

Polyamides offer another fine example of competition between the two giants of the chemical industry in the 1930s, IG Farben in Germany and E.I. Du Pont de Nemours & Co. in the United States. It is usually thought that the first polyamide was Nylon, developed between 1931 and 1935 by Wallace Hume Carothers of Du Pont. As a fiber, it met with legendary commercial success in the United States in the form of nylon stockings. What is less well known is that Dr Paul Schlack of IG Farben had also synthesized a polyamide, possibly even a few months earlier, marketed as Perlon but largely kept for military use. So during World War II, German and American parachutes were made of the same material.

The full range of processes involved in transforming synthetic materials into useful finished objects is too large to cover in detail here. This section will, therefore, concentrate on those that, given the cost of molds, tooling, and the production process itself, are adapted to industrial-scale manufacturing and mass production.

The injection-molding process involves injecting material in a fluid state into a mold at high pressure: from 300–2,000 bar (4,350–29,000 psi). All thermoplastics can be injected, as can some thermoset polymers and some polyester compounds. The technique can be used to mold any solid shape, and the surface quality can be finely

NYLON PARACHUTE FROM WORLD WAR II. 22,600 TONS OF NYLON WERE USED BY THE ALLIED FORCES, IN THE FORM OF PARACHUTES AND CHAFER FABRIC FOR TIRES.

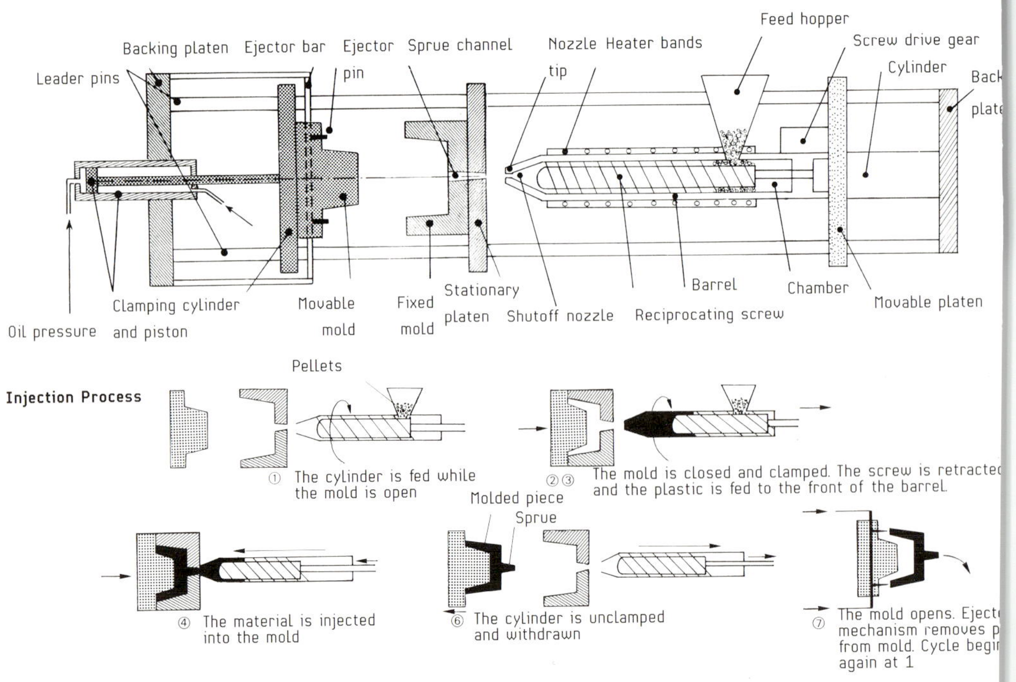
Feed hopper
Backing platen
Ejector bar
Ejector pin
Sprue channel
Nozzle tip
Heater bands
Screw drive gear
Cylinder
Back plat
Leader pins
Clamping cylinder and piston
Oil pressure
Movable mold
Fixed mold
Stationary platen
Shutoff nozzle
Barrel
Reciprocating screw
Chamber
Movable platen
Injection Process
Pellets
① The cylinder is fed while the mold is open
②③ The mold is closed and clamped. The screw is retracted and the plastic is fed to the front of the barrel.
Molded piece
Sprue
④ The material is injected into the mold
⑥ The cylinder is unclamped and withdrawn
⑦ The mold opens. Eject mechanism removes p from mold. Cycle begi again at 1

controlled by polishing, brushing, or sand-blasting the mold beforehand. The material can be volume-dyed, to any color imaginable.

This technique gave rise to a number of cult design objects after the war: Braun's beautifully shaped and boxed appliances in the 1960s, for example, molded in ABS. Then there are the objects that change materials over the years, adapting to the capability of new substances: for example, Joe Colombo's Universale chair, molded in ABS in 1965 and ten years later molded in polypropylene. Today, Philippe Starck's La Marie and Louis Ghost chairs are also made of injected polycarbonate.

In injection blow molding, material is first injected around a core in a preform mold. The core and preform plastic are then transferred to a second mold that has the outside dimensions of the finished object. The preform is reheated, and inflated to take the shape of the mold. This is a case of technology transfer from the glass industry. Plastic bottles and toys are made in this way, as well as parts for the automobile industry, isothermal units, and all manner of hollow shapes. The material used has to be thermoplastic. To refer to the table above, the materials used in this process are PET, PVC, SAN, ABS, PC, and PMMA.

In extrusion, thermoplastic material is pushed through a profiled die. The extruded material has a constant axial cross-section and adopts its final physical characteristics when it cools. The pieces obtained in this way can be as simple as a tube, or highly complex profiles used in the building or transport industries. It is also possible to coat metals or textiles in this way (the sheathing on electric cables is a common example), and to perform co-extrusions of more than one material at a time.

With extrusion blow molding, a film of "melt" is extruded as a cylinder, which is then sandwiched between the two halves of the mold, and inflated from the inside. As with injection blowing, only thermoplastics can be molded in this way: PE, PP, PVC, PS, SAN, ABS, and PC.

Simple, everyday containers, such as flasks, bottles, and containers, are obtained by extrusion blow molding. Blown film is obtained in a similar way: a wide cylinder of melt is blown up into a tall balloon, stretching and cooling to a very thin film.

facing page

top

AN INJECTION MOLD OPENS AND EJECTS FINISHED PIECES.

bottom

INJECTION MOLDING OF SYNTHETIC THERMOPLASTIC MATERIALS.

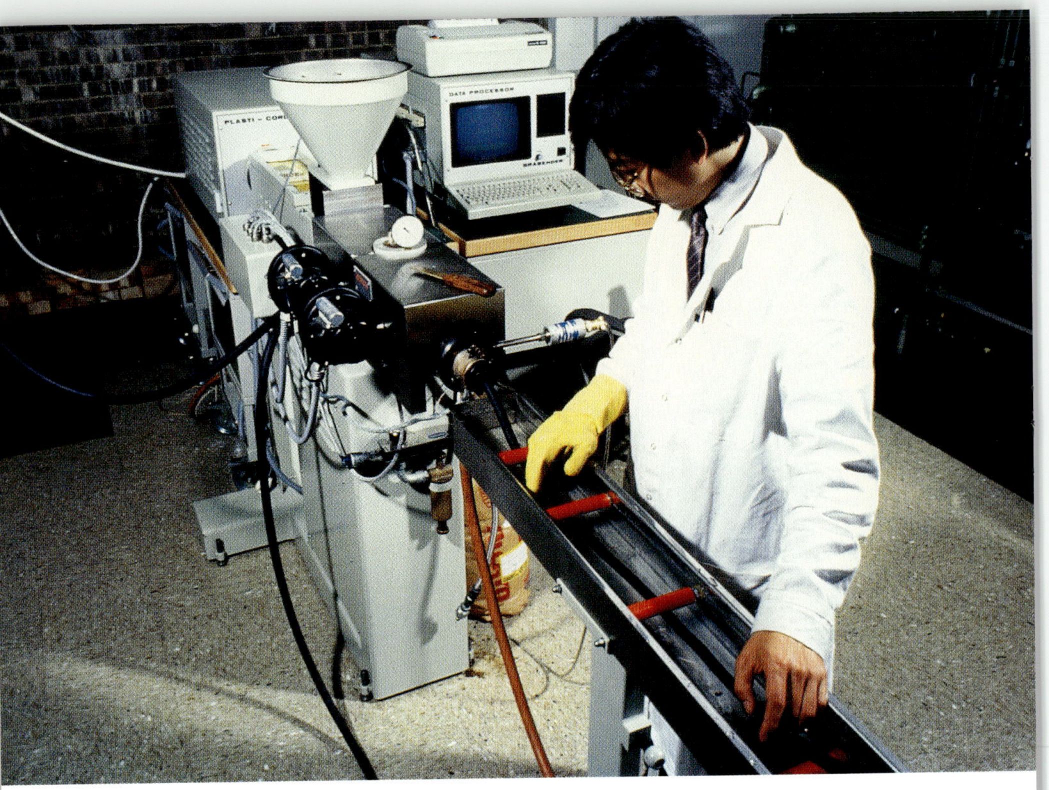

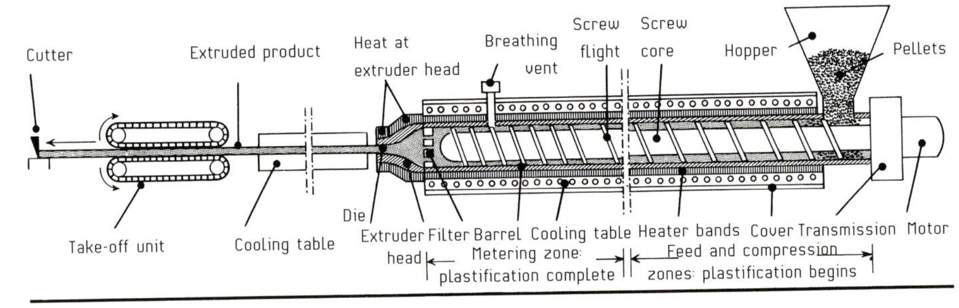

Cutter
Extruded product
Heat at extruder head
Breathing vent
Screw flight
Screw core
Hopper
Pellets
Take-off unit
Cooling table
Die
Extruder head
Filter
Barrel
Cooling table
Heater bands
Cover
Transmission
Motor
Metering zone: plastification complete
Feed and compression zones: plastification begins

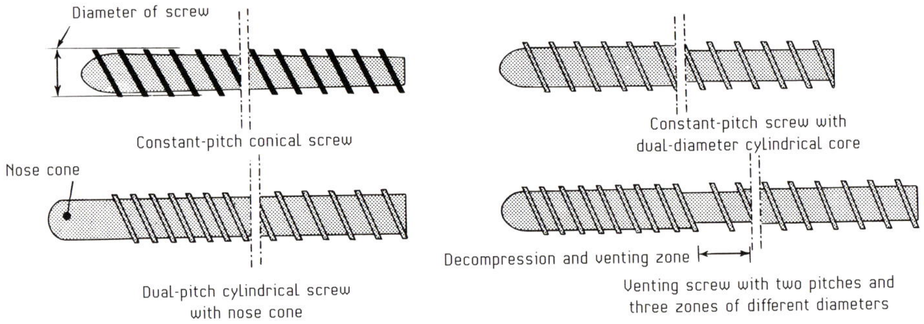

Diameter of screw
Constant-pitch conical screw
Constant-pitch screw with dual-diameter cylindrical core
Nose cone
Decompression and venting zone
Dual-pitch cylindrical screw with nose cone
Venting screw with two pitches and three zones of different diameters

above

BOTTLES MADE BY EXTRUSION BLOW MOLDING, ON THEIR WAY TO BE FILLED.

facing page

top

PROFILED SYNTHETIC THERMOPLASTIC PRODUCT EMERGING FROM AN EXTRUSION PRESS.

center

DIAGRAM OF A PROFILE EXTRUSION LINE FOR SYNTHETIC THERMOPLASTIC MATERIAL.

bottom

A SELECTION OF FEED SCREWS

OBTAINING A VERY FINE FILM
BY BLOWING A CYLINDRICAL
EXTRUSION.

facing page
MOLD FOR THE EXTRUSION BLOW
MOLDING OF POLYETHYLENE
BOTTLES.

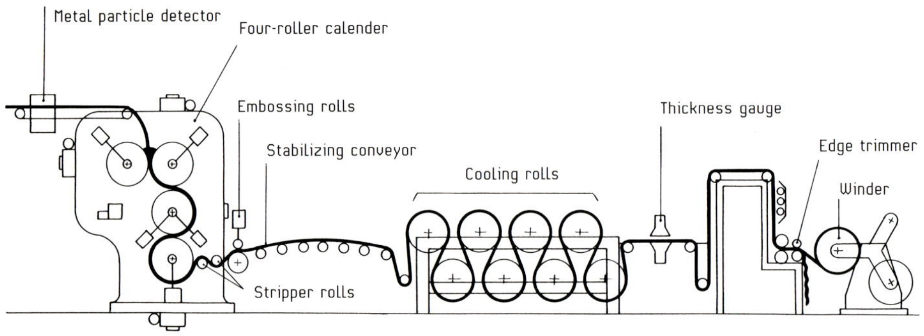
Metal particle detector
Four-roller calender
Embossing rolls
Stabilizing conveyor
Cooling rolls
Thickness gauge
Edge trimmer
Winder
Stripper rolls

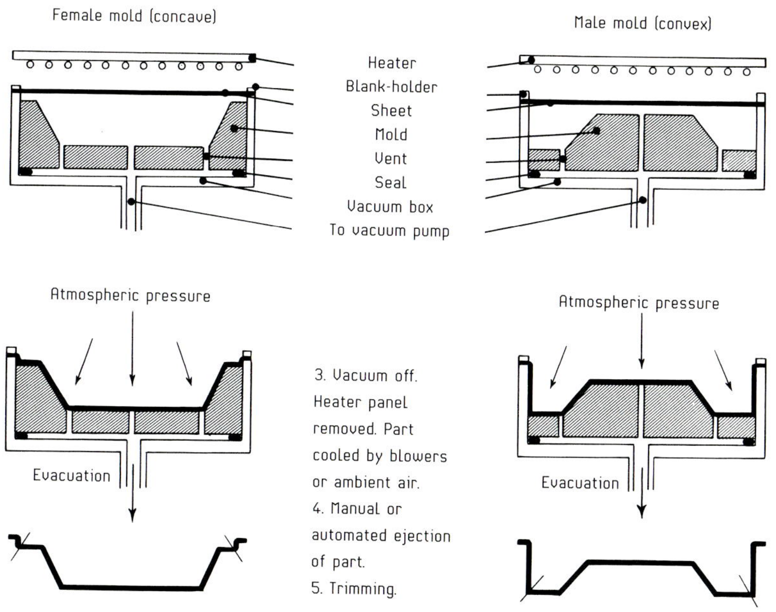

1. Heating. Sheet softens. Vacuum is off.

2. Vacuum is switched on. Softened sheet is sucked onto the mold.

3. Vacuum off. Heater panel removed. Part cooled by blowers or ambient air.
4. Manual or automated ejection of part.
5. Trimming.

Calendering is the passing of a material through a series of rollers. It is used to obtain rigid sheets, soft sheets, and film of thermoplastic materials: PS, SAN, ABS, PP, PVC, PC, and PMMA. The technique is mostly used to produce semi-finished products, particularly for thermoforming.

In thermoforming, a sheet is held in place in a frame, heated (generally by infra-red), and shaped around a mold by vacuum forming. From chocolate box packing to car bodywork, to an entire 20-foot boat with a dual skin hull and ABS cabin; more or less any continuous shape that can be removed from the mold can be made by thermoforming.

Rotational molding (or rotomolding) requires thermoplastic powder or prepolymerized liquid resins (PE, PP, PVC plastisol, PA, fluorinated resins, polyester) to be added to a mold that is then sealed. The mold is rotated around two axes and heated, thus distributing the material and coating the inside surface of the mold. The finished piece is hollow and can be fairly small (dolls' heads, toys, balls), medium-sized (fuel tanks for cars, grease traps), or very large indeed (canoes, toilets for building sites, portable site buildings). This technique was adopted by both Ron Arad and Philippe Starck, the former for his Little Albert polypropylene chair for Moroso in 2000 and the latter for his Bubble Club polyethylene chair and sofa for Kartell in 2001.

Compression molding is used for shaping phenolic and amino resins, rubbers, and glass fiber or preaccelerated polyester compounds, which harden as they undergo an irreversible chemical reaction at the end of the operation. This technique was used for molding Bakelite tableware, light fittings, and radio and television casings at the beginning of the twentieth century. Bakelite objects are now highly prized by collectors. Some, such as the DBH 1001 telephone designed in 1930 by Jean Heiberg for Ericsson or the 1939 Phonola radio designed by Livio and Pier Giacomo Castiglioni, are some of the great icons in the history of design. Today, compression molding has now largely been replaced by transfer molding, which is comparable to the injection molding of thermoplastics.

Simple casting is used for resins with one or two components that are in a liquid state. The material (polyurethane, polyester, epoxide, or silicone) is poured into a mold with one or more cavities. The technique is used for

preceding pages

left, top

CALENDERING PRESS.

left, bottom

DIAGRAM OF A CALENDERING LINE FOR PLASTIC SHEET AND FILM.

right, top

THE CITROEN *MÉHARI*, DESIGNED BY JEAN-LOUIS BARRAULT AND THE SEAB (SOCIÉTÉ D'EXPLOITATION ET D'APPLICATION DES BREVETS). THE FIRST VEHICLE WITH BODYWORK ENTIRELY MADE FROM THERMOFORMED ABS PANELS. CITROËN, FRANCE, 1968.

right, bottom

DIAGRAM OF VARIOUS THERMOFORMING PROCESSES FOR SYNTHETIC THERMOPLASTIC SHEET AND FILM.

facing page

REMOVING A ROTOMOLDED POLYETHYLENE TANK FROM ITS MOLD.

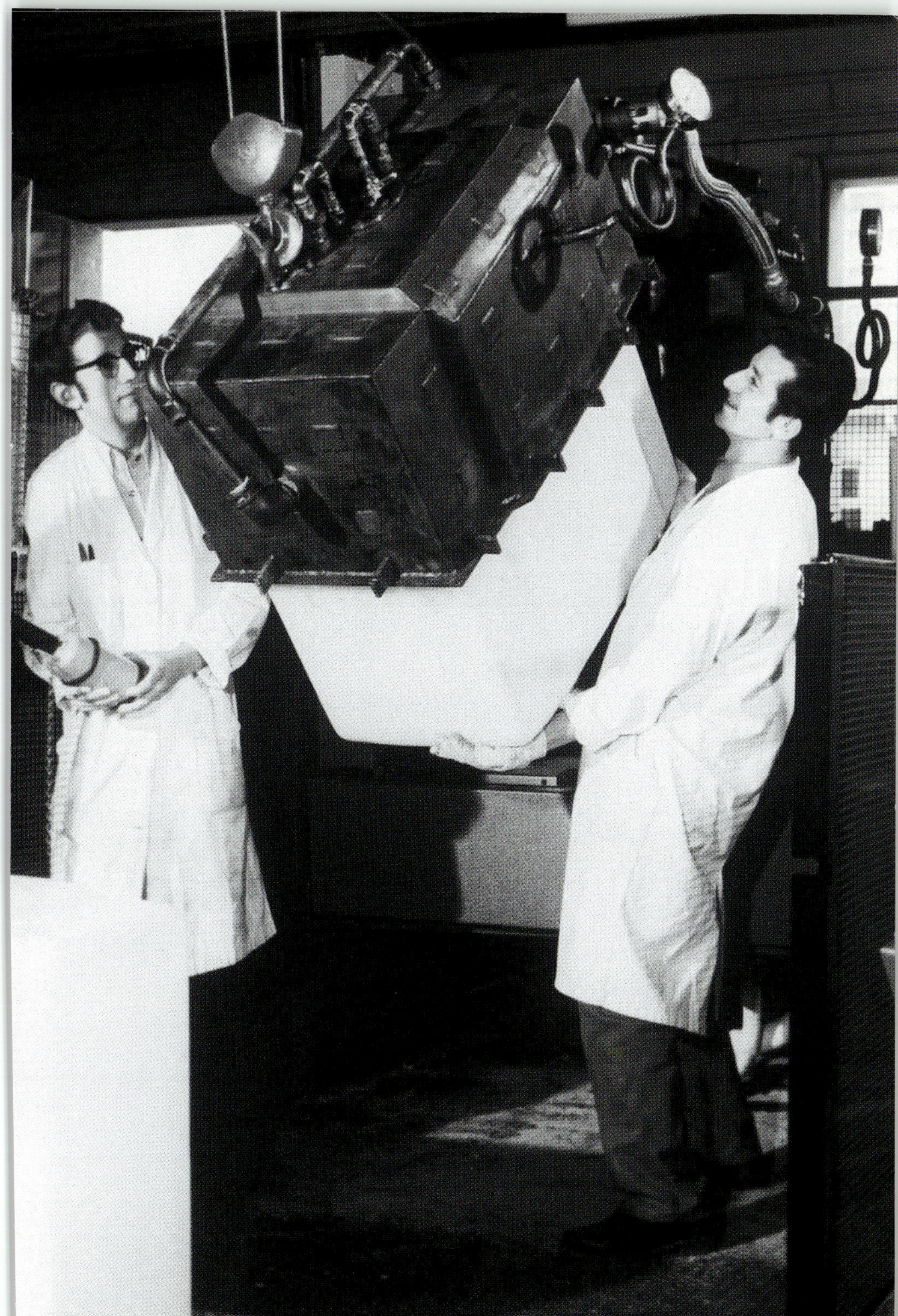

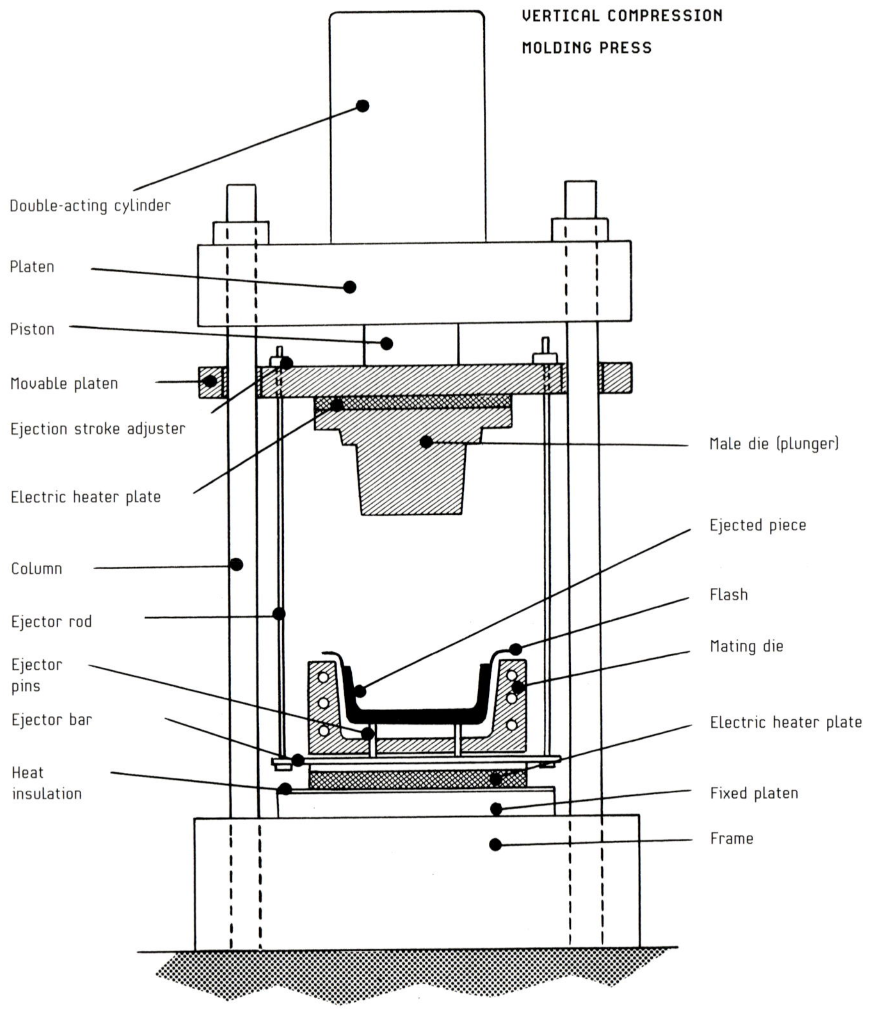

COMPRESSION MOLDING PRESS AS USED WITH THERMOSETTING PHENOLIC AND AMINO RESINS.

facing page
VERTICAL COMPRESSION MOLDING PRESS.

above

LIVIO AND **PIER GIACOMO CASTIGLIONI**, WITH **LUIGI CACCIA DOMINIONI**, RADIO, PHONOLA, ITALY, 1939.

left

BAKELITE RADIO SET PRODUCED BY FADA, UNITED STATES, 1934.

making wheels for skateboards, roller-skates, and shopping carts. It is also used for impregnating electronic components, making special technical parts, etc.

Dip coating can be used to protect an object with an elastomer or enhance its ergonomic qualities: safety gloves for industry or gardening, for example, or tool handles. The same technique is used for making condoms and other medical products. Coating has as many applications in the industrial domain (e.g. tarpaulins) and in fashion (coated materials for bags, clothes, cases, etc.). A layer of plastisol (PVC) or polyurethane is applied to a base material, which may be a woven textile, a knitted fabric, paper, cardboard, leather, or wood.

Contact molding is the most basic technique for shaping polyester and epoxide resins. It is used in industry, in arts and crafts, and in model-making, with very different goals, of course. After making a mold, a series of layers of resin and strengthening fibers is laid or sprayed into the cavity. Glass fiber can be used, either woven or not, as can carbon fiber or Kevlar. If layers of resin-impregnated fiberglass are used, any bubbles must be carefully removed at each stage. As it is difficult to change the material's color significantly, the cavity can first be painted with a special resin, known as "gel coat."

From just after World War II until the 1980s, fiberglass-reinforced polyester was a constant source of inspiration for designers, thanks to its remarkable mechanical strength and ability to assume complex forms. There is only room to mention a few of the many first-rate examples here: Eero Saarinen's and Charles Eames's fiberglass chairs in the 1950s for Knoll and Herman Miller respectively; Joe Colombo's Elda lounge chair for Comfort in 1963, Marc Held's bed for Prisunic in 1969, and the anthropomorphically styled Felt chair designed by Marc Newson in 1994 for Capellini.

Combining as it does resins and a reinforcing mat, fiberglass also belongs to a family of materials that is undergoing constant growth, to the benefit of technology in general: composites.

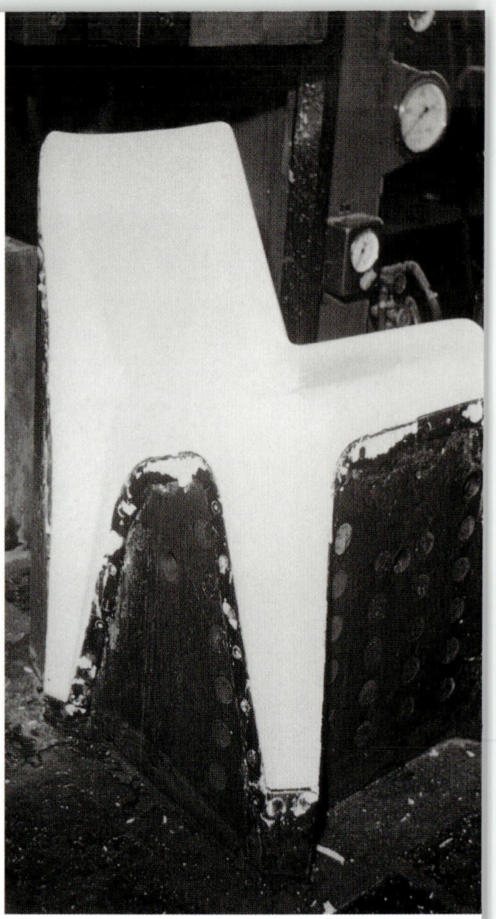

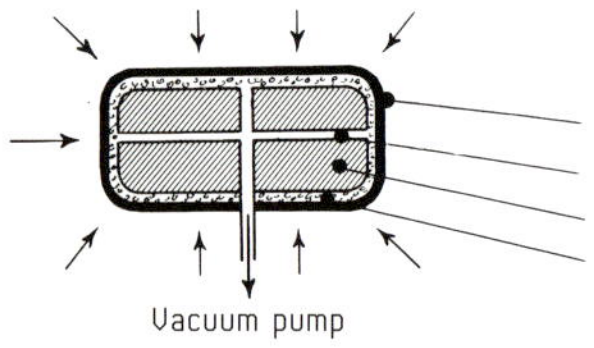

Bag molding
Bag (rubber or other flexible material)
Vents
Mold
Laminate

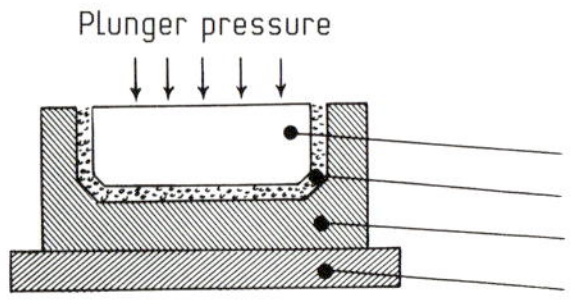

Punch molding

Synthetic rubber p
Laminate
Female mold
Base plate

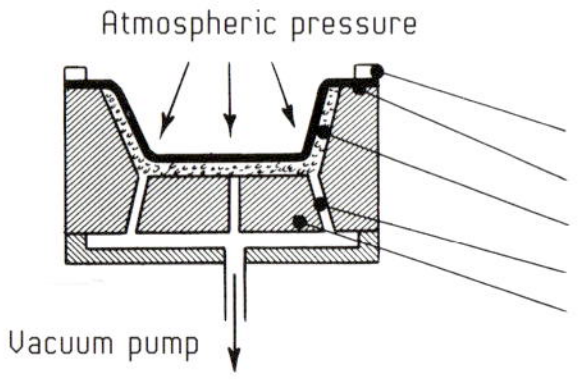

Resin infusion
Seal
Rubber membrane
Laminate
Vents
Mold

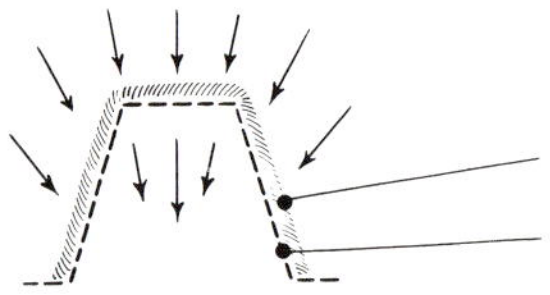

Screen preformin
Resin is sprayed c
drawn through the
preform by a vacu
Perforated screen

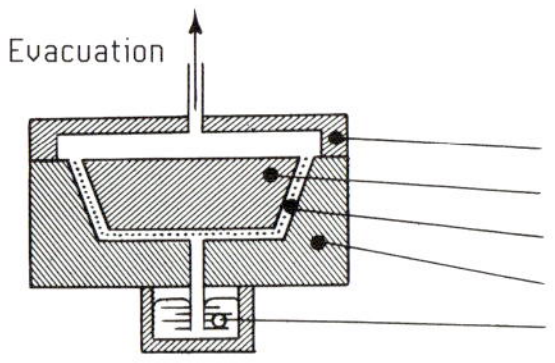

Vacuum injection molding
Vacuum box
Counter-mold
Dry fibers
Mold
Liquid resin

facing page

top

MAKING *CHAIR NO. BA 1171*

bottom

DIAGRAMS OF THE PROCESSES USED IN INDUSTRY FOR THE LOW-PRESSURE MOLDING OF COMPOSITES SUCH AS FIBERGLASS, WHERE A FIBER MAT IS IMPREGNATED WITH THERMOSETTING POLYESTER OR EPOXY RESIN.

HELMUT BÄTZNER, *CHAIR NO. BA 1171*, THE FIRST MASS-PRODUCED CHAIR MOLDED IN ONE PIECE FROM SYNTHETIC MATERIALS (FIBERGLASS-REINFORCED POLYESTER). IT MET WITH GREAT SUCCESS AT THE COLOGNE FAIR IN 1966. PRODUCED BY BOFINGER, GERMANY, 1964–1966.

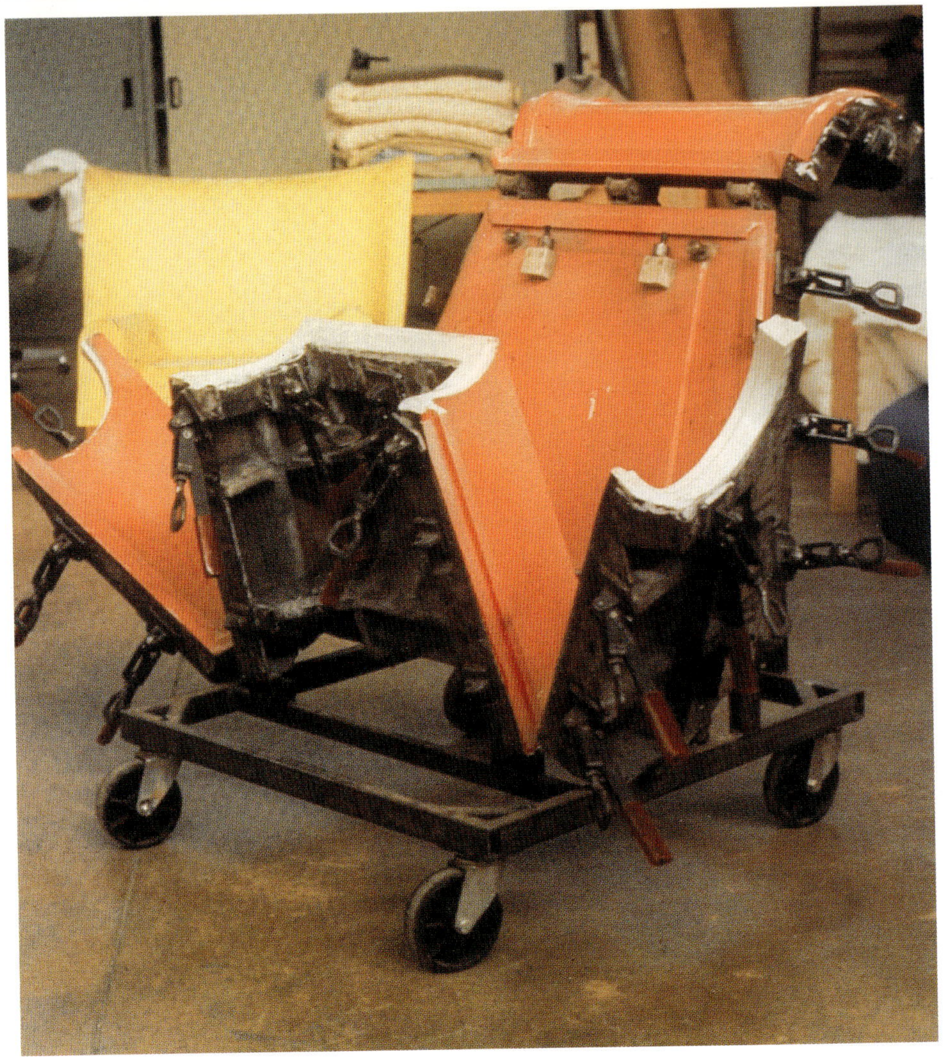

PAOLO DEGANELLO,

TORSO CHAIR, WITH DIFFERENT DENSITIES OF POLYURETHANE FOAM FOR THE SEAT, THE BACK AND THE END CUSHION. CASSINA, 1982.

facing page

CONSTRUCTION OF THE FIBERGLASS AND POLYESTER RESIN MOLD FOR THE POLYURETHANE FOAM BACK CUSHION.

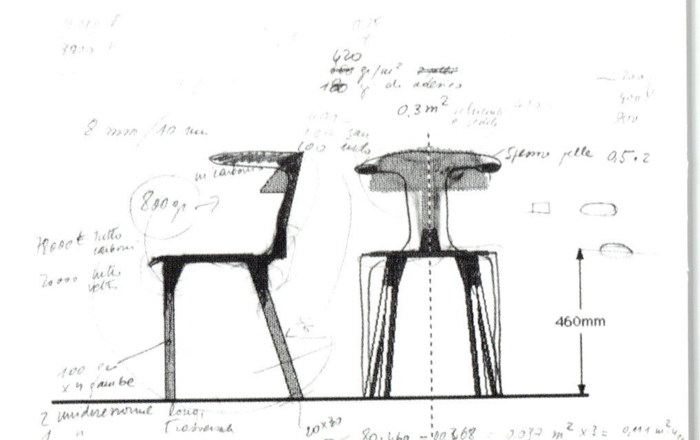
460mm

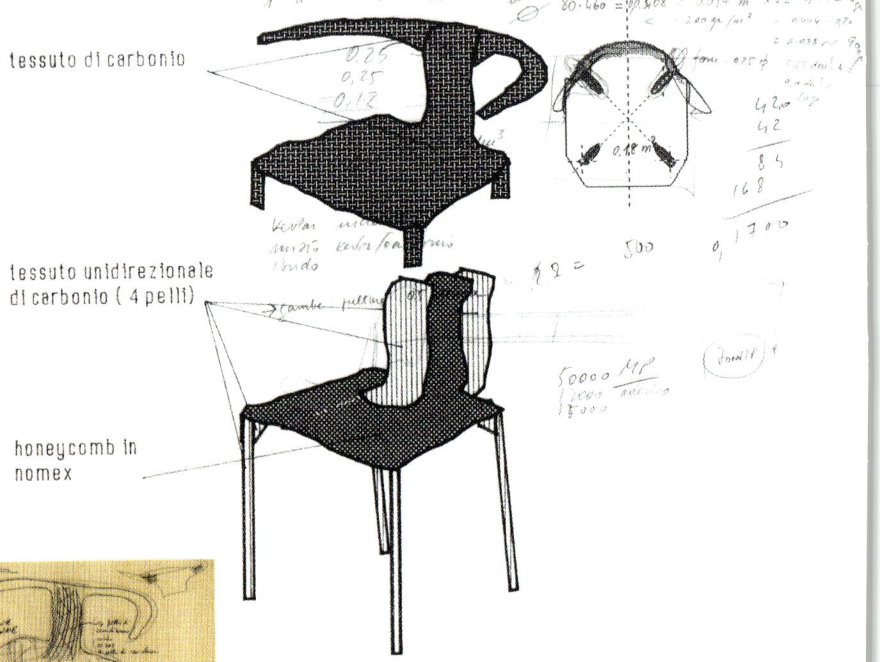
tessuto di carbonio
tessuto unidirezionale di carbonio (4 pelli)
honeycomb in nomex

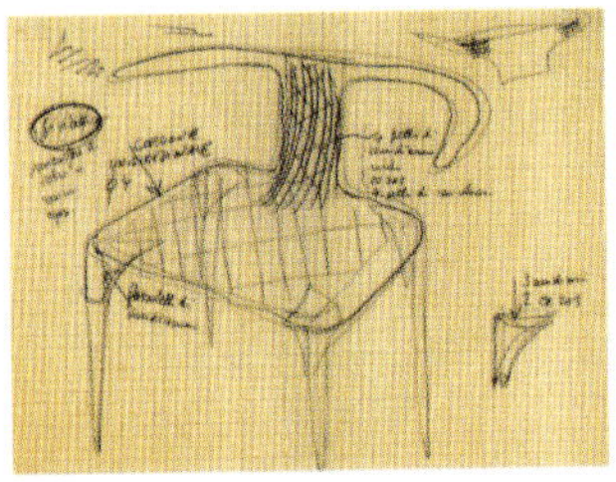

COMPOSITE MATERIALS

All the materials in this essentially diverse group are the product of combining several elements with complementary qualities, resulting in high-performance materials with very specific applications.

In many cases, composites are made by embedding fibers that have a high tensile strength in a substance (usually synthetic) that binds them together. The binding matrix can be thermoplastic (polypropylene, polyamide), but more often it is a thermoset substance (polyester, epoxide, phenolic resin, silicone, polyurethane, etc.). The substrate can also be mineral (carbon, silicon carbide), when great strength or heat resistance are required. It can even be metallic: both aluminum and titanium alloys are used. The fibers are as fine as possible—between 5–15 microns in diameter—allowing the finished piece to have sharper curves, increasing the breaking strain, and allowing the material, in some cases, to be processed as a textile. The fibers most often used are filaments of spun glass, sometimes woven into a mat. An alternative is carbon fiber, which is obtained by oxidizing acrylic filaments under heat (pyrolysis) then heating them to 2700°F in a nitrogen atmosphere. Kevlar (an "aramid", or aromatic polyamide) is also used, as is silicon carbide.

As well as by contact molding, composites can be shaped by injection molding. A molding press is used to inject a mixture of thermoplastic or thermoset resin and small fragments of fiber. This allows the cycle to be automated, and more complex forms can be produced at a faster rate. Fiber-reinforced polyurethane foam can also be injected, to make strong pieces of larger dimensions, such as car dashboard assemblies.

When the mixture is injected, the short fibers tend to align: a part can therefore be given increased strength in a particular direction. Even more strength can be achieved by introducing long fibers in line with the strongest traction forces the final piece will undergo. This has led to advanced techniques where fibers are mechanically aligned in layers. Filament winding is used to make pipes and hollow vessels such as storage tanks, which have to withstand very high levels of mechanical stress. The process is not limited to axis-symmetric designs: more complex shapes, such as helicopter blades, can also be wound.

ALBERTO MEDA, *LIGHT LIGHT* CHAIR, MADE BY ALIAS, ITALY, 1987. DETAIL OF VARIOUS PARTS USED IN THE MANUFACTURING PROCESS.

As well as molding fiber-reinforced composites, it is also possible to extrude profiles or sheets, which can then be shaped at a later stage. These materials complement another important group of composites: sandwich composites. These are often made on demand to very specific dimensions, but they are best known to the public as sheets and boards. A sandwich composite material (from which we expect mechanical performance) is obtained by welding or bonding one sheet of material, chosen for its tensile strength and rigidity, to another which is weaker but has complementary qualities, such as lightness or good insulation. A wide variety of top sheets can be used, such as metal, plywood, or a laminate. The lightweight core may be composed of an expanded material, or a corrugated sheet of metal or laminate, or a cellular structure made of aluminum, cardboard, or even Kevlar. Aluminum honeycomb panel, for example, has been used for aircraft floors. When a sandwich panel has only to be load bearing and decorative, melamine-coated particleboard is suitable and is widely commercially available.

Designers and industry alike were quick to seize on the possibilities of shaping glass-fiber reinforced polyester, from the 1940s onward. In the last few years designers have witnessed the technical achievements of carbon fiber and have begun to experiment with it in a similar way. Sylvain Dubuisson's Table Composite and Alberto Meda's Light Light chair (both 1987) are milestones in this technology, and it was only the prohibitively high production costs that prevented both designs from being commercial successes.

HELICOPTER ROTOR MADE
OF COMPOSITE MATERIAL.
FILAMENTS HAVE BEEN WOUND
ONTO THE SURFACE, ORIENTED
FOR MAXIMUM RESISTANCE
TO THE MECHANICAL STRESSES
THE BLADE HAS TO UNDERGO.

above and facing page

RON ARAD, *OH VOID I*,

CARBON FIBER AND RESIN

CHAISE LONGUE, 2002.

POLYMER NANOCOMPOSITES

Some of today's cutting-edge technology takes place on a very small scale indeed: nanotechnology deals in microns (a millionth of a meter). In space research, weight and volume must be kept to an absolute minimum to enable probes to travel further and further. Medical research has started to work on little vehicles that can explore the entire blood system.

Nanotechnology is now being applied at the atomic level, which, thanks to scanning tunneling microscopes, can be enlarged one hundred million times and observed. At this unlikely scale, composite materials are benefiting from research such as that done at the Canadian Conseil National de Recherches (CNRC).

Nanotechnology is recognized as one of the most promising directions for technological developments in the twenty-first century. Where materials are concerned, nanocomposite ceramics and polymers are growing areas of multidisciplinary research.

Polymer nanocomposites are polymers reinforced by small quantities (under 5 percent) of nanoscopic particles of clay. The mineral improves the mechanical and thermal resistance of the base polymer considerably:
• fire retardant and barrier properties are improved
• strength is improved without significant increase in density
• optical properties and recyclability can be improved.

A WORD ABOUT STOCK-REMOVAL TECHNIQUES

Throughout this overview, much has been said about techniques of machining materials that minimize waste. This is indicative of the strong economic demands of an expanding market, and their influence on the development of technology. In the decades since World War II, the industry has, to a great extent, managed to eliminate the reworking of mass-produced items. The constituent parts of a household appliance, whether made of synthetic material or aluminum alloy, are ready for assembly as soon as they leave the injection-molding press.

STABILIZED ALUMINUM FOAM. VERY LOW DENSITY CELLULAR MATERIAL WITH GOOD RESISTANCE TO COMPRESSION, MAKING IT IDEAL FOR LIGHTWEIGHT PANELS. ITS ESTHETIC QUALITIES ALSO MAKE IT AN INTERESTING CHOICE FOR INTERIOR ARCHITECTURE. MADE BY ALUSION, CANADA.

When some finishing is needed—an automobile engine block, for example—the piece will at least be moved on to the appropriate station by an automated process.

There are areas where shaping by stock-removal is inevitable—the machining of the very tools used for mass production, for example: injection molds, extrusion dies, stamping dies, drawing punches, die cavities, etc. Computers are used to help design and to automate the machining process of these parts, but the tools used are primarily the most ordinary ones: drills, cutters, lathe tools, broaching tools, grinding wheels, reamers, and boring bars.

When the metal is too hard or the required shape too intricate for the piece to be machined in the normal way, modern erosion techniques may also be used to cut complex shapes out of metal. Chemical milling involves masking parts of the surface and subjecting the exposed areas to corrosive acid. In electrochemical machining (ECM), the workpiece is submerged in an electrolyte, and a current is passed between it and a pre-shaped tool. The metal is dissolved until the shape of the tool (the cathode) is reproduced in negative in the workpiece (the anode).

Electrical discharge machining (EDM)—sometimes called "spark machining"—uses an electrode to produce a rapid series of high-frequency electrical discharges to remove particles of matter from the workpiece, which is submerged in a dielectric fluid. Several variations exist, including "plunge" or "sink" EDM, which uses a shaped electrode; wire EDM, which uses a brass wire; and the use of a rotating EDM tool to make holes or mill contours.

Some sort of machining—even if it is an advanced technique like laser-cutting, water-jet cutting, or erosion milling—is indispensable at some point in the manufacture of a product. Computer-aided design and manufacturing is used for rapid prototyping, and give us a glimpse of a future in which items could be made without having to be extracted from a mold, or a pair of stamping dies, or a press. We no longer need to carve mold cavities out of steel globes, yet the machine that does it for us holds a laser-cutting head and heating nozzle, which were made by other machines or production robots, equipped in turn with classic tools for trimming, milling, turning, drilling, boring, and grinding.

PRODUCTION LINE FOR AUTOMOBILE CHASSIS AND REAR AXLES, CITROËN PLANT, PARIS, 1931.

RENAULT PLANT, BOULOGNE-BILLANCOURT, 1952. TRANSFER MACHINES ("ROBOTS") ARE BEING USED.

CUTTING ENGINE BLOCKS ON
AN INGERSOLL CONTINUOUS
MILLING MACHINE, GUTENBERG
PLANT, RENAULT, 1931.

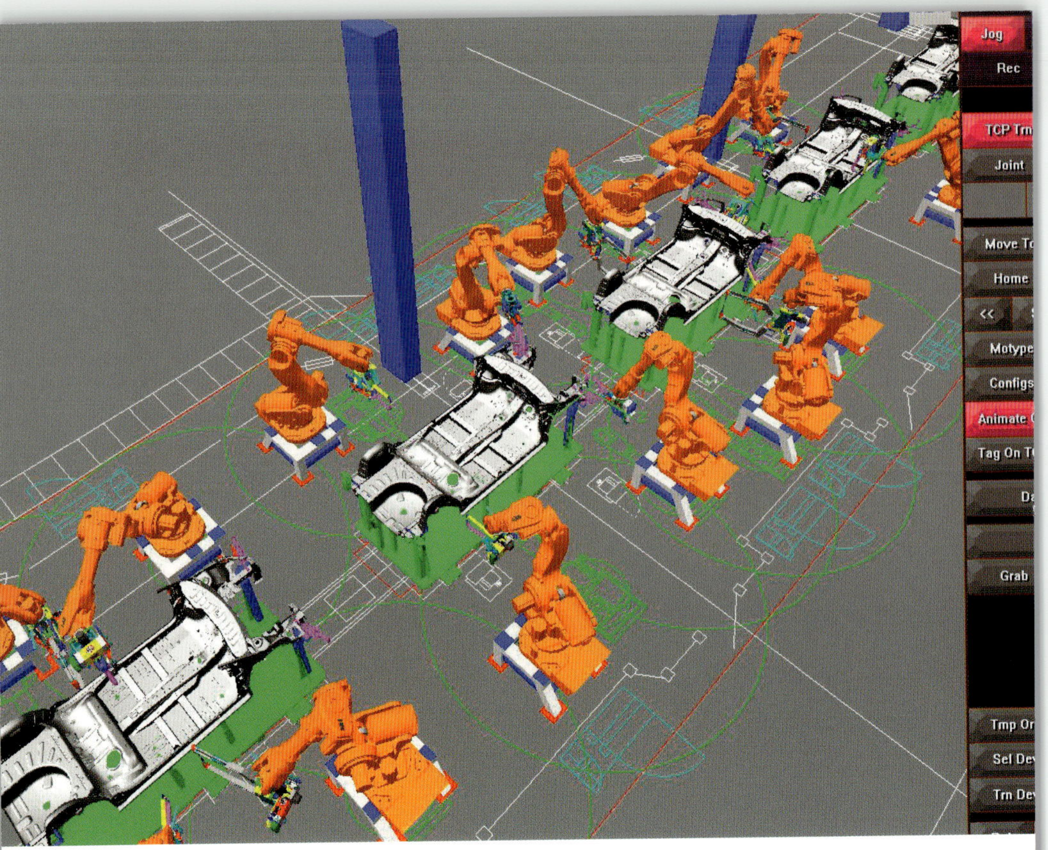

SIMULATION OF MANUFACTURING ROBOTS AND THEIR MOVEMENTS ON A CAR PRODUCTION LINE, SOFTWARE BY DASSAULT SYSTÈMES.

It looks very much as though the techniques mentioned in these pages are those to which designers, engineers, and technicians alike will continue to turn, for many years to come, when creating the objects of the future.

SO WHICH MATERIALS SHOULD WE CHOOSE?

The choice of materials must be dictated by the specific characteristics of the product in question. When mechanical strength is important, we can turn to metals, composites, polymer concrete, or certain wood derivatives. If, for example, the choice falls to metal, one automatically thinks of possible surface treatments to protect it from corrosion, or possibly to change its appearance and give the material a different "meaning" thanks to its new shape and color.

If the function to be fulfilled is not a structural one, strength is of secondary importance, and the choice will be made on the desired finish, or other physical, optical, thermal, or dielectric characteristics.

Where optical constraints are concerned, the choice between transparent synthetic materials and glass is one of refractive index, with PMMA or PC plastics being more refractive than glass. Synthetic materials would seem the sensible choice, unless scratch resistance is likely to be a problem.

Where thermal insulation is an issue, the best solution is a vacuum (as with a thermos flask). Failing that, cellular synthetics are a very interesting compromise: they not only insulate well, but can be incorporated in a sandwich structure for good resistance to bending stress. If dielectric constraints apply, all synthetic materials are good insulators.

Still other factors need to be taken into account: dimensional constraints are a deciding factor in the choice of material and transformation processes. The size of the production run also has a direct influence on the return on investment: the larger the run, the smaller the cost of investment to be absorbed by each item at sale, and the more feasible it is for the designer's imagination to be allowed a free reign.

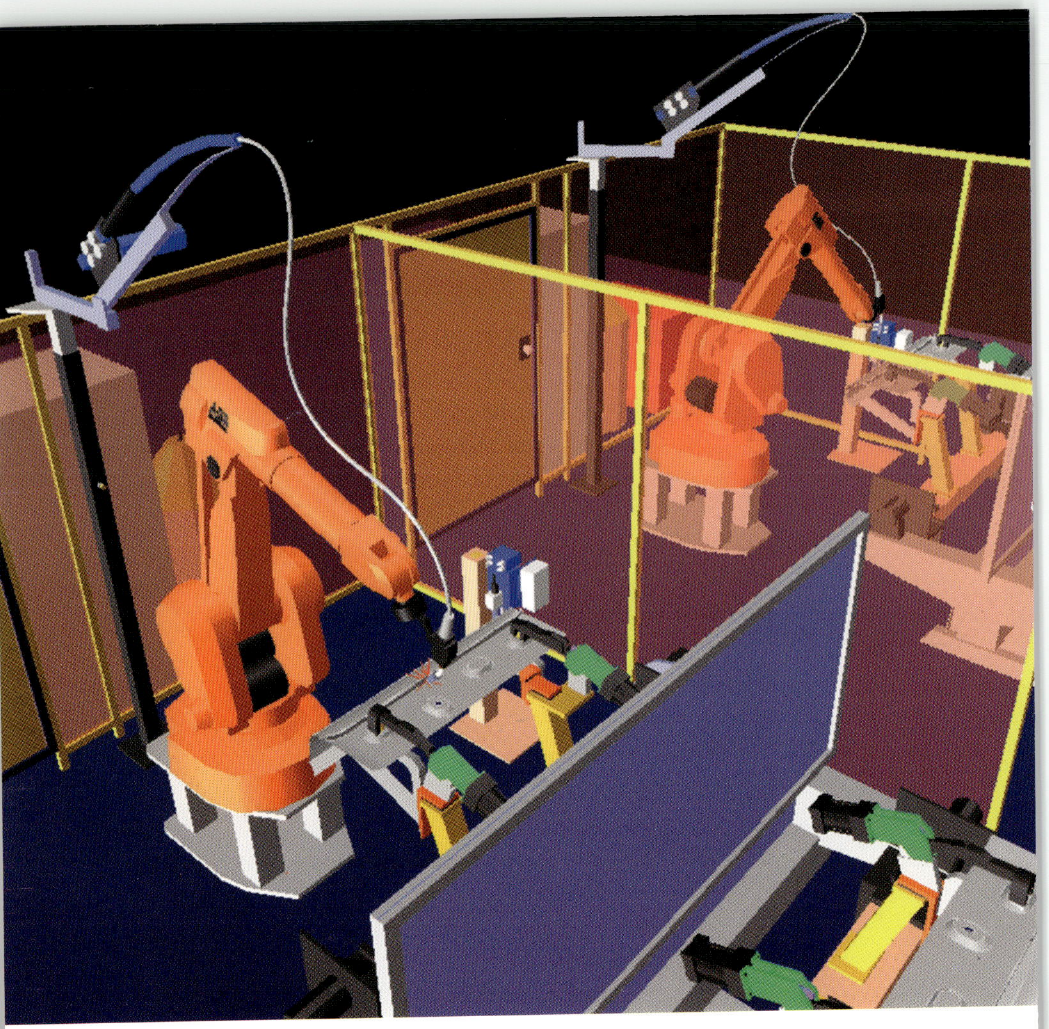

VIRTUAL FITTING OF FACTORY ROBOTS, DASSAULT SYSTÈMES.

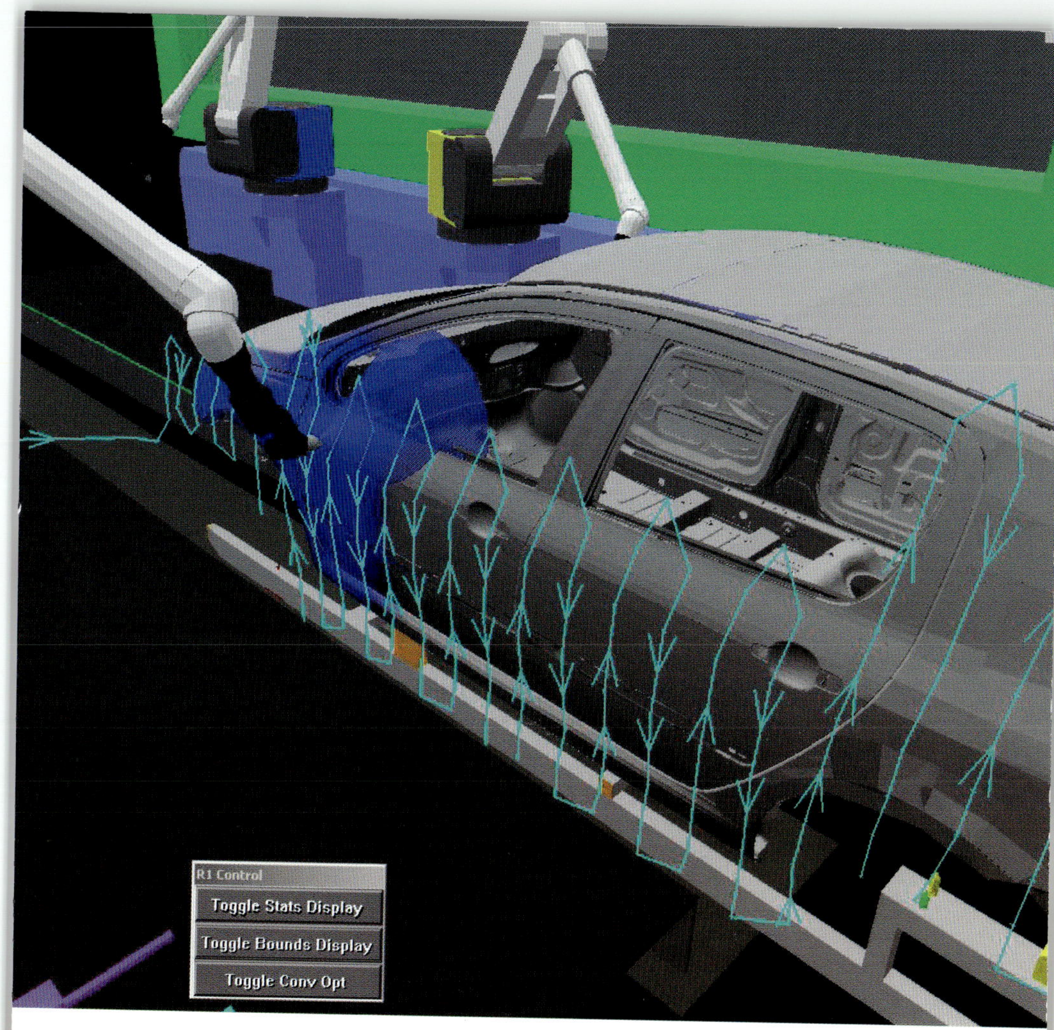

PLANNING THE TRAJECTORY
OF ROBOTIC PAINT-SPRAYER ARMS,
VIRTUAL FACTORY SOFTWARE
BY DASSAULT SYSTÈMES.

CHARMILLES

TRUMPF
259459

Transformation technology and machining techniques have been evolving extremely quickly, and computer-based tools have completely transformed certain processes. The limitations now lie with the designer's creativity. Everything depends on the designer's ability to find original solutions to the changing habits, needs, and demands of the consumer.

preceding pages

left, top

EDM, "SPARK MACHINING": THE ELECTRODE HOLDER IS LOWERED INTO THE DIELECTRIC LIQUID TO MEET THE WORKPIECE, HELD IN PLACE AT THE BOTTOM OF THE TANK.

left, bottom

ELECTRICAL DISCHARGE MACHINING.

right, top

DIGITALLY-CONTROLLED LASER CUTTING MACHINE, ISSINOX, ISSY-LES-MOULINEAUX, FRANCE.

right, bottom

TOOLING OF A THERMOPLASTIC INJECTION MOLD.

facing page

FRANÇOIS AZAMBOURG, *YVETTE* LUMINOUS WEAVE LAMP IN PMMA (POLYMETHYLMETHACRYLATE) OPTICAL FIBER, AWARDED AN "APPEL PERMANENT" FROM VIA (VALORIZATION OF INNOVATION IN FURNISHING), FRANCE, 2001.

DIGITAL MATERIAL

JEAN-BAPTISTE TOUCHARD

It all happened with so little fuss in the space of a generation, without anyone really noticing. And yet, you just have to look at the earliest Bézier curves, at Martin Newell's Melitta teapot modeled in 1975—over thirty years ago now—and at the acts of the first SIGGRAPH, to see how far we've come. For a few thousand dollars, designers can now possess a tool of phenomenal power—or, more precisely, an entire palette of complementary, interacting tools.

The calculator developed into a typewriter, then became a drafting machine, then a machine for trying out product appearance, transparency, and texture, then a tool for photographic or video assembly, or for inlaying still or animated multimedia scripts. By a mix of graphics and number-crunching, a mechanism can be simulated, an operation animated, and a mechanical resistance analyzed.

The digital model lends itself to all and any transformation. Set up on screen, one software component passes information on to another, each fulfilling its task. The shell for the next coffee-grinder can be drawn using a CAD program[1] in the Rhodes agency, and then transmitted in an instant to a rapid prototyping machine in the factory in Nanjing that will produce the mold.

The calculator has also become a communication tool. Where once it was isolated, today a workstation can share the contents of its disks with every other computer on the planet. It has become one of the innumerable components of the Internet that designers can delve into to retrieve information, draw up a budget or a business plan, and manage methodology. Telephones and personal digital assistants synchronize contact lists and appointments; videos and cameras are digital. Documents are drawn up, altered, and circulated using software.

preceding page

MICROPROCESSOR

facing page

MARTIN NEWELL TEAPOT MODELED WITH DISCREET 3DS MAX SOFTWARE. NEXT TO THE TEAPOT, A 1950S SERVICE IN MELAMINE, SIGNED MÉLAWARE.

1 CAD: Computer-aided Design. This notion incorporates all the techniques and tools, essentially 2D and 3D technical drawing, that allow an object to be represented.

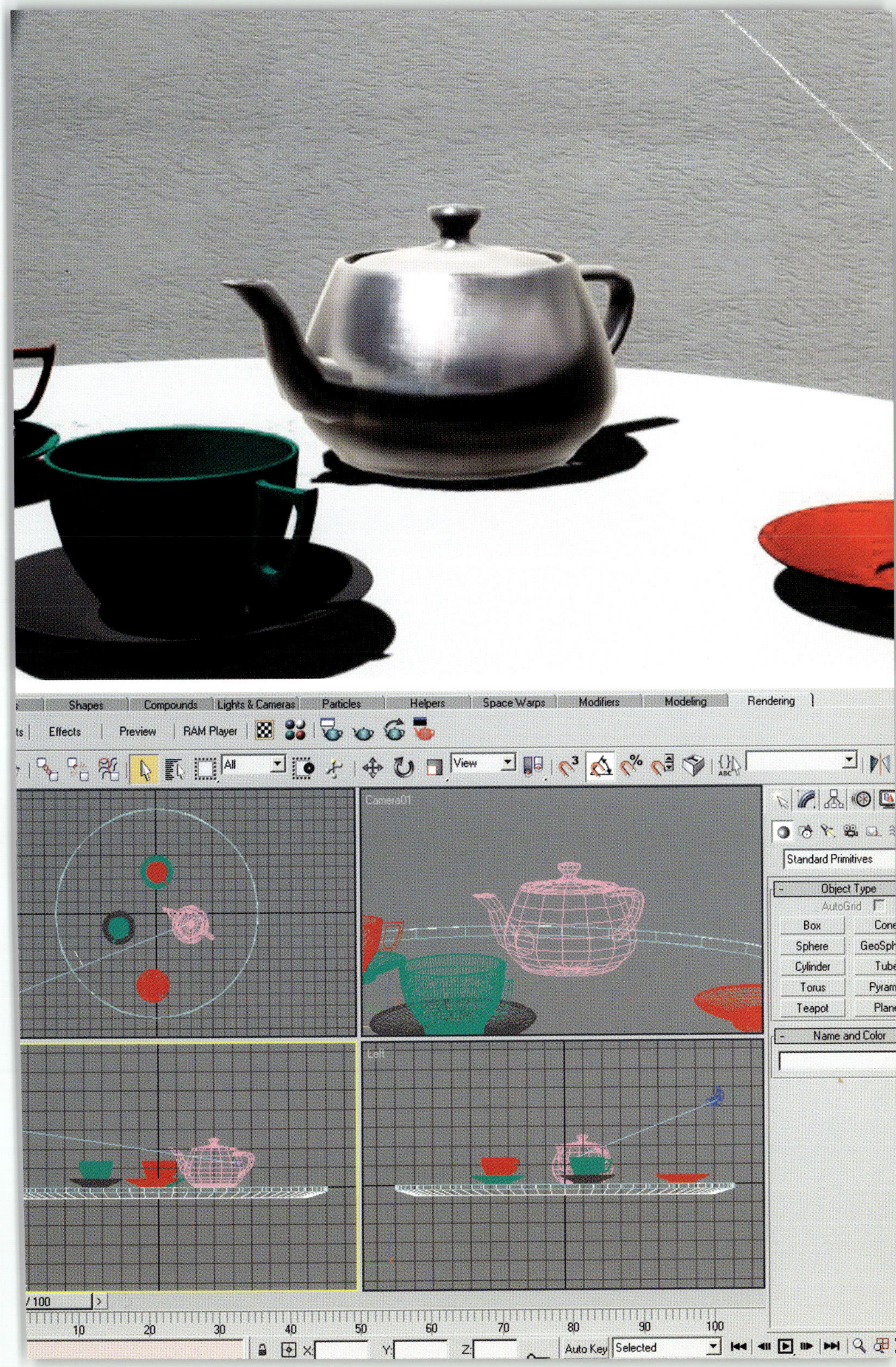
Shapes
Compounds
Lights & Cameras
Particles
Helpers
Space Warps
Modifiers
Modeling
Rendering
Effects
Preview
RAM Player
All
View
Camera01
Left
Standard Primitives
Object Type
AutoGrid
Box
Sphere
Cylinder
Torus
Teapot
Name and Color
Auto Key
Selected

FRACTAL SURFACE IN ALTUGLAS, MACHINED BY PAUL GEEVERS IN 1985 TO A PROGRAM BY JEAN-BAPTISTE TOUCHARD UNDER THE AUSPICES OF THE RESEARCH ON TEXTURE UNDERTAKEN AT THE ÉCOLE NATIONALE SUPÉRIEURE DE CRÉATION INDUSTRIELLE.

THE COMPUTER IN THE SERVICE OF THE DESIGNER

The process of design has become virtual, but that is not to say it has become dematerialized. The dictionary defines "virtual" as something "which only exists putatively; potential, possible...." In spite of the "only," a pejorative hangover from a past that respected the concrete, it is a definition rooted in an intention and, as such, is very close to the design-intention that employs drawing as a medium.

The virtual is not the antithesis of the real. It is located upstream. If need be, it can be materialized in an instant; files from CAD are accessed directly to produce machining templates that will control the CAM[2] robots.

Today's machines include the digitally controlled milling-machine whose computer-controlled step drive has replaced the displacement levels on table and cutter. It can sequence thousands of movements, remove the toughest materials in successive "passes" fractions of a millimeter deep, or automatically change cutter to fine-tune some minute detail. However, such machining is not invariably faster than manual work executed by a talented lathe- and milling-cutter operator, especially if one factors in the time taken to write the program controlling the movements of the milling-machine and for optimizing the machining templates. But its incredible precision encourages daring machinings of a complexity unheard of with manual methods. In the 1980s, Isabelle Bonnet, an early graduate of the prestigious École Nationale Supérieure de Création Industrielle in Paris, redesigned a teapot on which the analysis of research on surface textures was undertaken using EUCLID, a CAD tool devised by Matra Datavision, while Paul Geevers explored Mandelbrot fractals to create entirely novel surface effects.

2 CAM: Computer-aided Manufacture. This concept includes the set of practices and tools employed to control a manufacturing robot. Digitally-controlled milling machines have since been succeeded by electroerosion machines, then by rapid prototyping machines using electropolymerization or selective sintering.

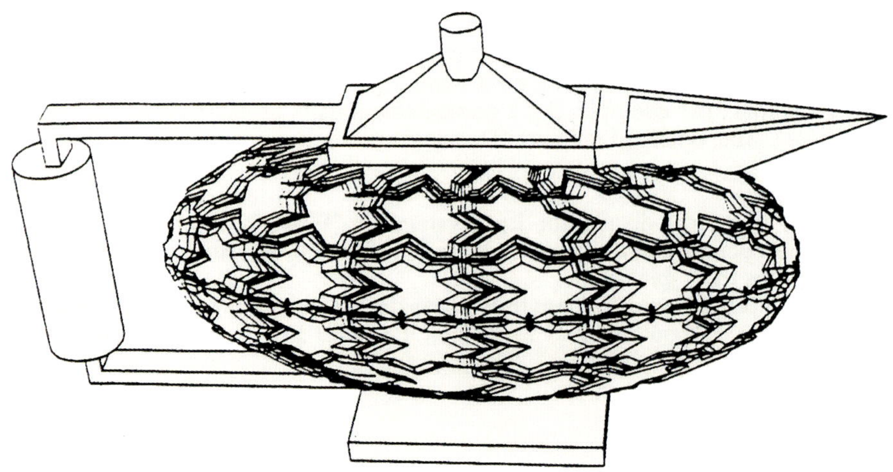

above and facing page

ISABELLE BONNET, CAD PROJECT FOR A TEAPOT AS A STUDY IN TEXTURED RELIEF.

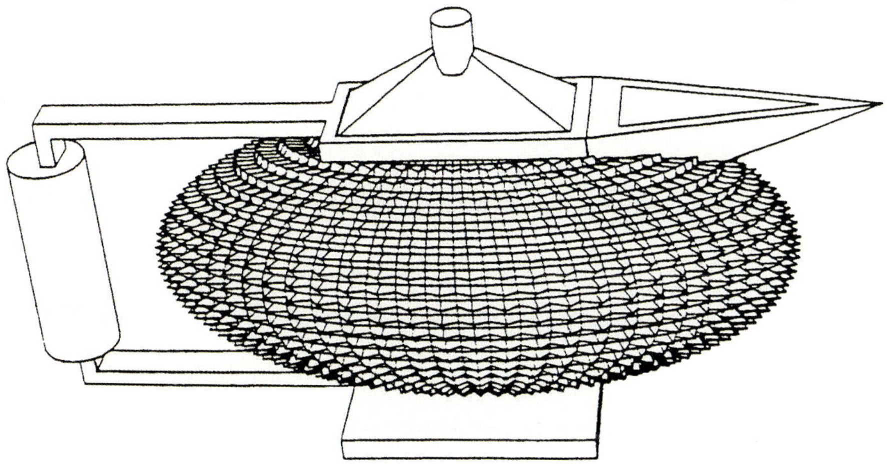

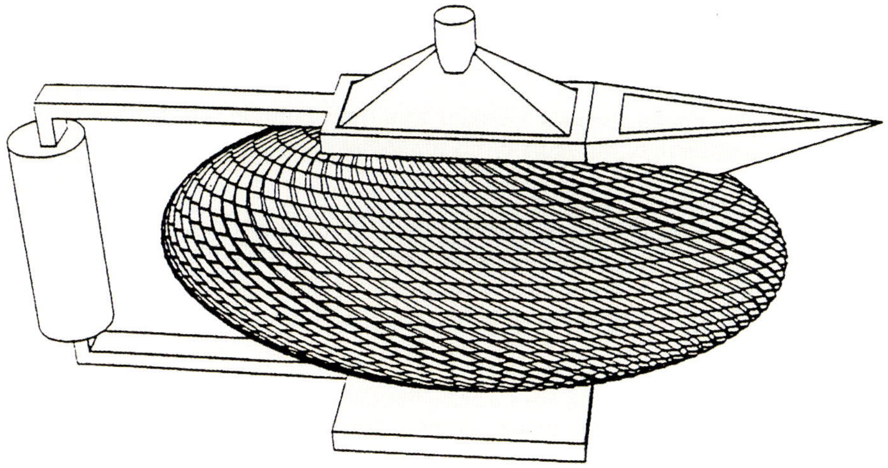

ROBOTIZED DRILLING OF FALCON AIRCRAFT WINGS, DASSAULT FACTORY, MARTIGNAC, FRANCE.

The digital milling-machine can produce complex, chaotic forms, but it can also attain an impeccable finish. At car shows around the end of the 1970s, it was fun spotting the first cars built using CAD tools. The small neon display lights would reflect curves regularly over the body, whereas molds built in the classic manner gave rise to more broken surfaces, which scatter light[3]. Today, not a single mass-produced automobile presents this type of irregularity. Turbine blades more than 13 ft. (4 m) tall, like those that do several hundred revolutions a minute in power stations on the Three Gorges in China, are now machined to the nearest micron, the result of intensive hydrodynamic calculations. The tiniest defect would set up vibrations that could result in catastrophe. The aircraft industry demands similar levels of accuracy: airflow is simulated, and each component is analyzed, tested for resistance, and optimized for weight. Insofar as it is impossible as yet to reduce the size of air or water molecules, these digital simulations are far more precise than simulations on models.

Of course, these singularly pure forms stem from the originator's first jottings; rooted in his intuition and culture, these provide the fundamental direction for the project, but subsequently these outlines will be corrected, refined, and checked by a whole host of calculations and computer simulations. Admittedly, digitally controlled machining is ill-adapted to industrial mass production. Painfully slow, it also produces too much waste material. It is perfect for designing a unique piece, a prototype, but is unsuitable for turning out hundreds of units a second on an assembly line. For that, recourse must be made to molding, thermoforming, or stamping. But that is no impediment; CAM is then employed to produce the injection molds or extrusion dies that are then deployed in plastic injection-molding, metallurgy, or glassmaking. Besides, before launching out into production, time would have been allocated to check the mold using rheological[4] simulations with parameters corresponding to the material injected. In this manner, one can forestall the risk of inadequate filling, check if the vents are correctly positioned, or gauge whether the piece produced presents

3 The term "spline," which appears in all 2D- and 3D-drawing software packages in association with the name of Pierre Bézier, refers to the way in which curves should behave to represent the connections between two panels riveted together.
4 Rheology is the theory describing the flow and deformations of materials under the action of constraints. The viscosity of the materials is the essential parameter for models used to demonstrate the rate of shear and other potential mechanical weaknesses in cast or drawn pieces.

ALSTOM
ALSTOM

facing page
TURBINE FRANCIS, 30 FT.
IN DIAMETER, MADE BY ALSTOM
AT LA CIOTAT FOR THE DAM
OF THE THREE GORGES
(PHOTO. E. LAMPERTI/ALSTOM).

above and following pages
VIRTUAL FACTORY SET UP
BY DASSAULT SYSTÈMES FOR
BUILDING THE FALCON 7X,
ALLOWING THE TEAMS SCATTERED
OVER VARIOUS PARTS OF
THE GLOBE (ARGENTEUIL AND
MÉRIGNAC, FRANCE, LITTLE ROCK,
ARKANSAS...) TO WORK ON
A VIRTUAL TERMINAL AND SHARE
A DATA BASE THAT DESCRIBES
ALL ASPECTS OF EACH PART OF
THE AIRCRAFT, ENABLING IT TO
BE EXAMINED IN EVERY DETAIL
AT EACH PRODUCTION SITE.

any structural weaknesses. The mold-maker's expertise remains paramount, but today he has on hand computing tools that enhance it considerably.

These processes, at one time the preserve of heavy industry and advanced technologies, are becoming increasingly accessible. Similar methods and tools are now deployed to realize prototypes and molds for small electrical appliances or luxury craft pieces. Designers armed with digital files enjoy far greater control over their creations, keeping them longer than they used to in the days when they had to hand their portfolios and models over to the engineering and design department or to R&D. At present, designers provide information that is implemented directly by the manufacturing units and is far more implicated in the adjustment and development phases of the industrial process than it used to be.

The most recent incarnation of CAM is three-dimensional printing, a rapid prototyping system using stereolithography or powder-sintering. Stereolithography is a photopolymerization process. A laser controlled by computer solidifies a liquid resin to a very fine thickness and the piece is built up in layers. The technique of sintering again makes use of a laser beam to solidify fine layers of a powdery material contained in a vat that is lowered and filled as the piece builds up. Once complete, the piece is cleaned and removed from the unpolymerized substance. It still takes hours to extract such a piece and the range of materials remains fairly limited, though each day brings new ones. In addition to the relatively simple polymers with which these techniques originated, manufacturers are now coming up with metallic materials and ceramics. These systems can generate forms that are impossible to manufacture by machining or molding: in the formal arena, they constitute a new departure compared to industry's traditional products.

If these techniques were to become applicable to industrial-scale production, it is easy to dream—as Laurent Lebot and Victor Massip have on their witty site, monsieurFaltazi.com—of a company that manufactures products only once they have been ordered over the Internet, ensuring delivery times similar to that of a stock-held product[5]. It is unnecessary to expand

5 Such solutions are already available for prototyping. One example is www.materialise.com, which proposes the realization of prototypes based on files sent by Internet in under 24 hours.

MARCEL DASSAULT AT

THE FIRST CATIA CONSOL

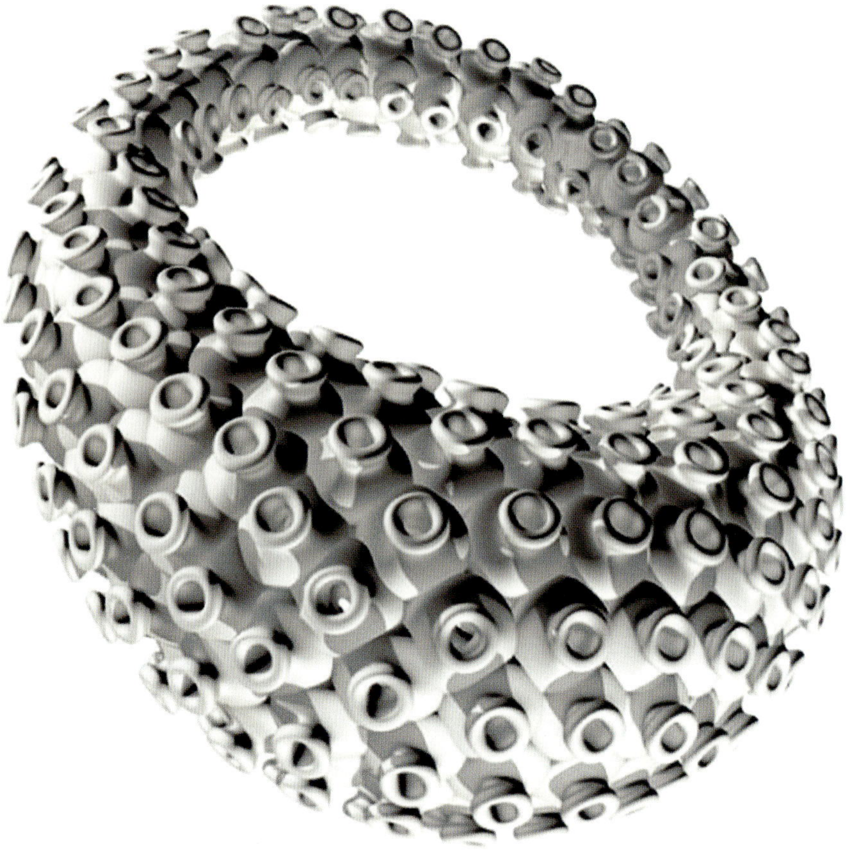

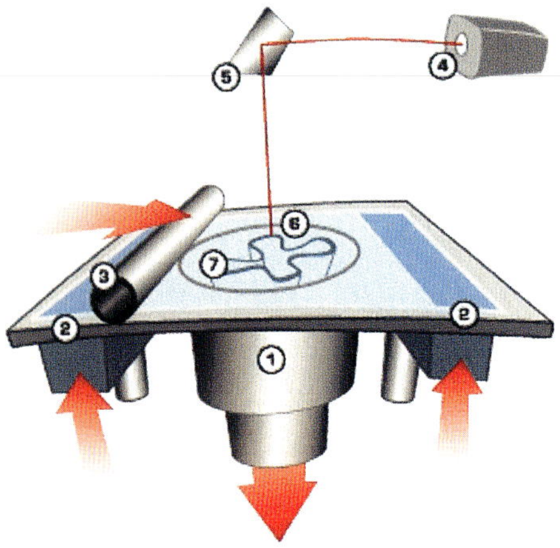

THE BAGOOZ HANDBAG,
BY MONSIEUR FALTAZI.

on the economic interest of such processes, but they give whole new meaning to the expressions "just-in-time production" and "zero stock."

Monsieur Faltazi proposes a new service to the general public: the instantaneous production of objects. Faltazian objects are manufactured by the technique of selective laser sintering, more commonly used for rapid prototyping (see diagram on facing page).

a - A piece of software breaks up a 3D CAD model into a set of basic 2D layers corresponding to transverse sections varying in thickness between .003" and .02" (76μm to 508μm). The choice of layer thickness has an impact on the precision and the surface finish, as well as on the time needed for manufacture.

b - The powder is preheated to below melting-point. Preheating is necessary to avoid the piece becoming deformed and to reduce the quantity of energy required for fusion.

c - The central platform [1] is positioned at the top of the tank.

d - In alternation, one of the two plates on the input tanks [2] is elevated to feed new material to the surface. A system of rollers [3] removes and pushes the powder across the surface from one side of the container to the other, building up a bed of powder.

e - The laser beam [4] focused at the surface of the resin is deviated vertically along the X and Y axes using a scanning mirror [5]. It thus draws a 2D geometric section [6] on the surface of the layer following a precisely defined template.

f - The lifting plate [1] supporting the object is lowered after each layer to a depth equivalent to the base layer.

g - The process continues layer after layer until the system finishes building the piece [7].

h - Submerged in the powder, the piece is brought to the surface and automatically cleaned.

But this dream can be pursued in quite another direction: perhaps the unique piece will no longer be the preserve of the art connoisseur, or confined to the production of bespoke furniture. In this scenario, Gaetano Pesce's diversified series or the production of unique pieces as produced by the Alchimia Studio or by the Memphis group would become viable in all industrial sectors. The theoretical propositions of the likes of Alessandro Mendini or Ron Arad, considered utopian until not long ago, are now perfectly realizable.

Jewelry-making benefits from an economical structure that fosters the use of such techniques. In 1992, François Scali called upon the services of CAD Design[6], an arm of the APCI, to execute the "plaster casts" for commemorative medals he had designed for the Winter Olympics at Albertville. Never before had the engineering departments of the Paris Mint seen work of such refinement by a designer. This series of medals was the starting-point for a technique that has since become almost banal. Pieces are now made with spectacular precision, with molds produced from prototypes realized by selective sintering.

The development of mechanisms or the use of materials such as shape-memory alloys (SMAs) combined with digital modeling allowed Thierry Holemans to make flower-jewels that open and shut with variations in temperature. One of the pioneers of digital imagery in the luxury goods sector, Louis Vuitton, is no sluggard either. Though design lead times have been slimmed down, the complexity of the objects and their manufacturing quality has soared. The traditional tools and skills of the craftsman cohabit harmoniously with the most advanced digital technologies.

The formal freedom that arises from these systems is all the more gratefully received as CAD software has improved in leaps and bounds. The time when straight lines and 90-degree angles gradually made room for Bézier curves and surfaces and for the omnipresent splines is a distant memory. A good 3D-modeling program today provides access to formal components that, instead of curbing the imagination, give it wings. NURBS[7] deformation modeling propose modifiers such as twist, stretch, flatten, melt, and overlay. Luigi Colani's heirs could rethink "biodesign" with a couple of clicks of a mouse.

Designers can even execute very precise models starting from a complex object, whether general view drawings or photographs. "Reverse engineering"[8] analyzes a shape or mechanism in order to determine its

6 This service, implemented in 1983 on the initiative of Anne Marie Boutin, president (CEO) of the Agence pour le Promotion de la Création Industrielle, by Jean Zetoun, pursuing facilities he had previously set up for architects, allowed designers access to the most powerful CAD tools available at the time. **7** NURBS: Non Uniform Rational B-Splines. **8** The idea of reverse engineering is often associated with industrial espionage. One "dismantles" a competing product and takes the results as a springboard. The idea presented here is a little different.

facing page
RON ARAD, BOUNCING VASES PROJECT, OBJECTS MADE USING RAPID PROTOTYPING. THE PICTURES SHOW THE VARIOUS STAGES IN THE C ANIMATION OF THE MODEL. BOTTOM

bottom, right
GEOFFSPHERE PRODUCED EMPLOYING THE SAME PROCESS.

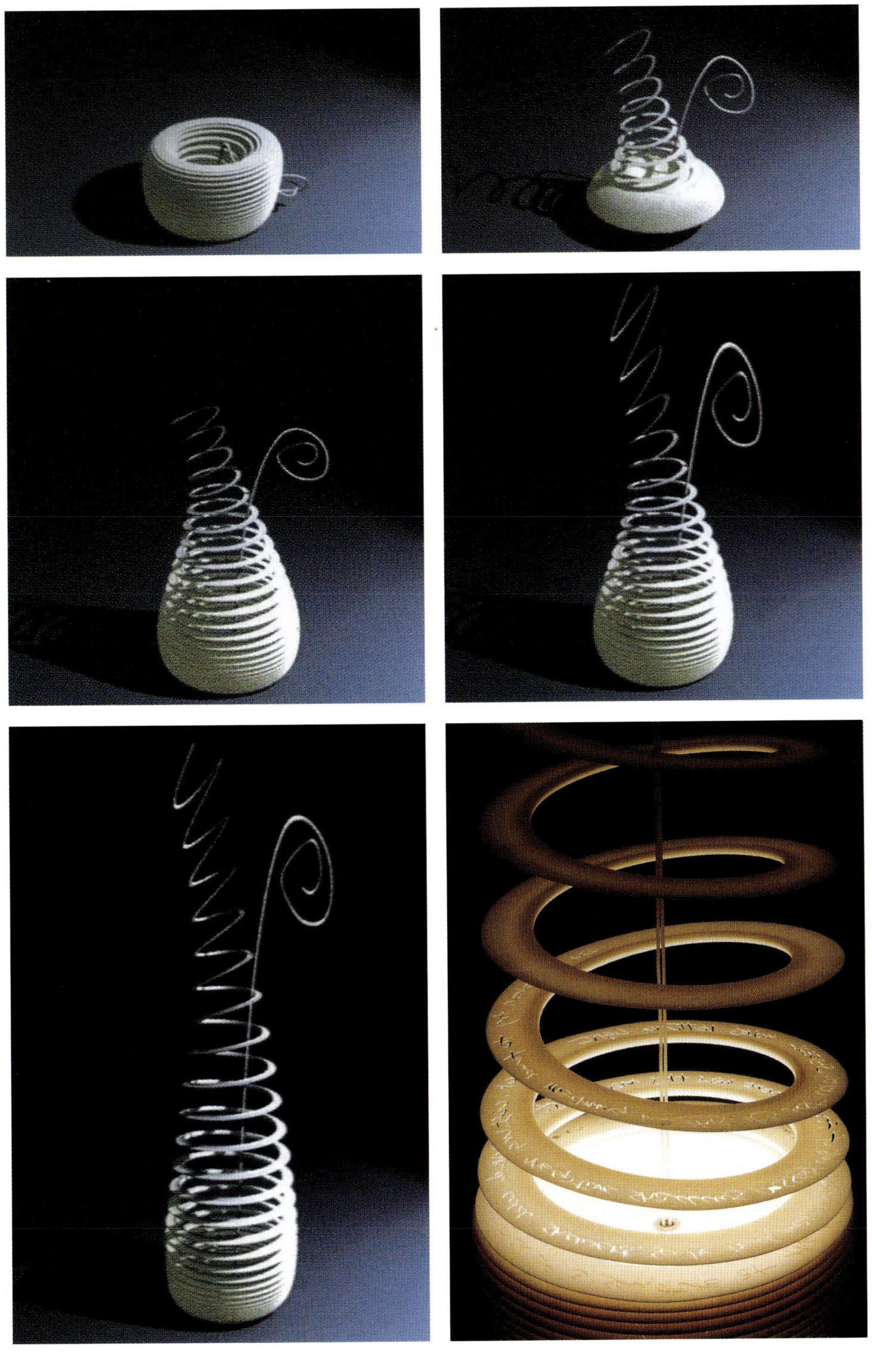

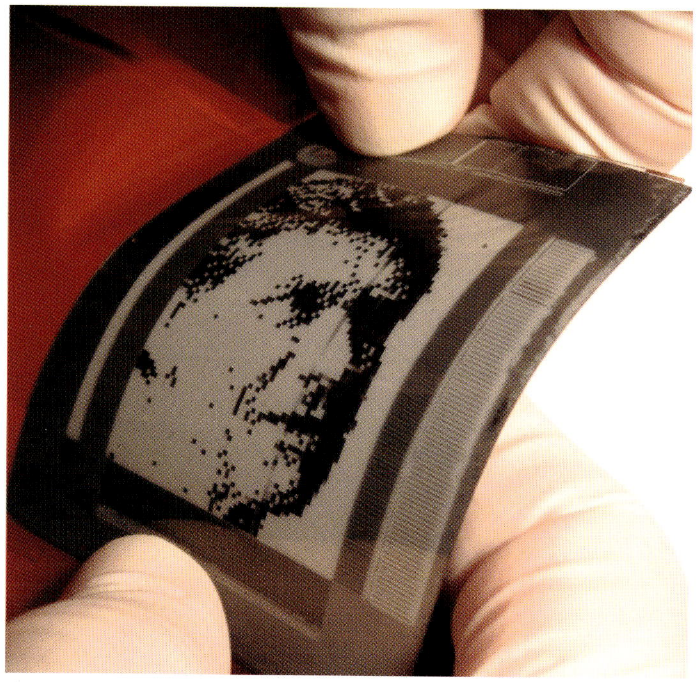

above

THE AS YET SMALL-SCALE, STATE-OF-THE-ART ULTRAFLAT SCREEN PROPOSED BY THE E INK CORPORATION IN JULY 2004.

below

ELECTRONIC INK MAKES IT POSSIBLE TO ENVISAGE ULTRAFLAT SCREENS THAT WILL FEEL SIMILAR TO PAPER. IN THIS MANNER, ONE MIGHT DESIGN AN ELECTRONIC BOOK EACH PAGE OF WHICH WOULD BE A SCREEN. MEMORIES AND ELECTRONICS BUILT INTO THE BINDING WOULD PROVIDE ACCESS TO AN ENORMOUS RANGE OF CONTENTS.

elementary components. This top-down approach is prevalent in the design of systems or software in which the designer starts with the end result and tries to determine its components with ever greater precision. The design process is thus the "reverse" of the traditional approach, which prefers to assemble simple forms and build them into complex constructions, rather like Lego.

Early CAD systems could harness only basic elements: segments or circles, parallelepipeds, spheres or cones, and Bézier curves or surfaces. Today they suit those who prefer more advanced design processes. Expertise and practice serve as guides in choosing among the countless methods available to achieve the desired result. Realistic rendering tools utilize photographic sensors to obtain information on the textures with which the model will be furnished.[9] Conceivably, in some laboratory or other, someone is developing "feelers" capable of acquiring other types of information on the object analyzed: its density and mechanical resistance, or even—why not—its internal structure as revealed by X-rays. A draft scribbled on a paper napkin would be immediately transmitted to the company's CAD system and, by the time the designer has finished his coffee, he'll find a comparatively advanced plan at his workstation; it will be enough for him to connect the (electronic) paper napkin to the computer in his office.

Utopian? Not really. It is precisely this that an industrial firm with a fondness for paper, such as Rank Xerox, has in mind in its investigations into electronic ink. Electronic ink technology relies on a paper on which each point oscillates between black and white, making each sheet into an extremely fine screen. One has to overcome the problem that reading on screen is uncomfortable or at least incompatible with the notion of reading as pleasure. An electronic book designed in this manner would comprise a large number of pages that the reader could flick through, thereby enabling an overall view which is normally impossible with on-screen reading. The featured text can be modified to the user's requirements, according to what appears in memories concealed in the binding. At a hotspot, or even via a mobile, readers could upload the latest novel before catching their train. Eventually, any notes made could also be saved for future consultation. Readers might not experience the sensory pleasure associated with reading a conventional book, but still the object would be more beguiling than the average lap-top.

9 For instance, Minolta has reaped the benefits of its expertise in the photography field with a 3D digitizer capable of simultaneously delineating geometry and texture.

THE DIGITAL IMAGE IN THE SEARCH FOR A CULTURAL IDENTITY

"Yes, but the image lacks warmth!" is a common complaint of the digital image. A synthesized image possesses neither the charm, nor the elegance, nor the strength of a fashion sketch, a photograph, or, to an even larger degree, a painting. Try as it might, the digital image has not yet managed to become an "eighth art" (after cinema). In the graphic arts, it is seen as tolerable—one step removed from technical drawing, on a par with illustration.

This debate, however, resembles the one that—once the provocative stances of the Impressionists and of abstract art had been taken on board—prevented photography from gaining artistic status before the end of the twentieth century or, more recently, the supposed opposition between film and video, before the latter broke free of TV reportage and addressed itself to the realm of storytelling. The digital image will have to bide its time before some of its productions attain the status of artworks. But here, too, things are advancing very fast. Very early on, advertising made use of images whose technical assets can endow their content with an impression of vigor and authority.

From 1970 on, artists such as Orso Martinelli made fake digital images, entirely carried out with an airbrush and, moreover, completely impossible to generate on the computers of the time. Similarly, IBM was quick to vaunt the quality of the CATIA system, showing a perfume bottle that appeared to emerge from a wireframe design produced in this manner.

From the very beginning of the 1970s, digital image designers were keen on animation, on the moving image. Peter Foldes produced animations[10] that herald the effects obtained by morphing. In the same period, Gerard Marinelli produced the credits for the French television program *Thalassa*, which still seem pretty fresh and which did much to raise the profile of computer art.

If computers in the early 1980s required hours of calculation to produce images that seem unsophisticated today, the

10 One thinks in particular of *Metadata* made in 1971 and *Hunger* made in 1974.

ELHO 06 05: AFTER MARCEL DUCHAMP. PROPOSITION FOR A BOOK COVER BY ORSO MARTINELLI, 1980.

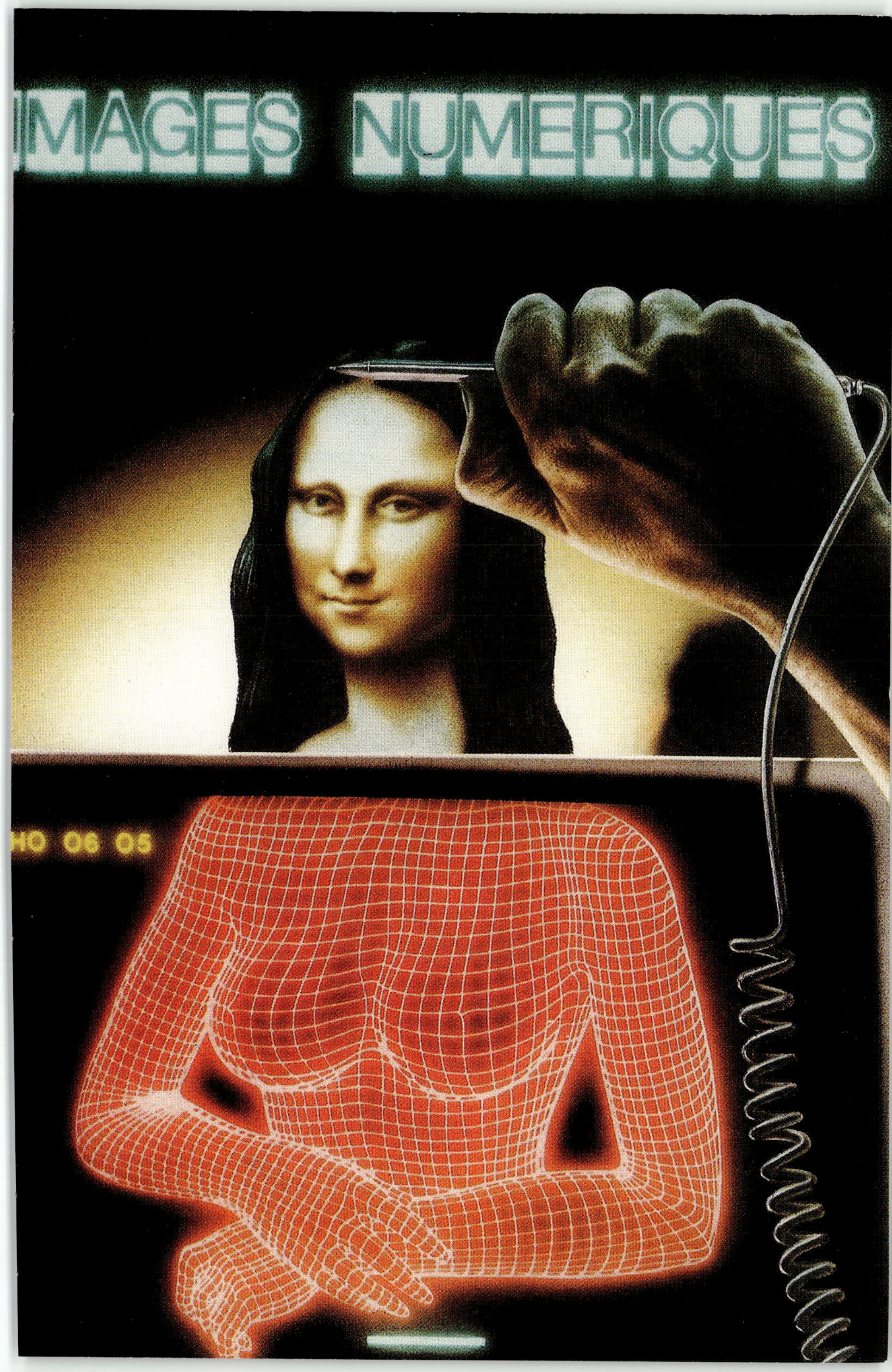
IMAGES NUMERIQUES
HO 06 05

LINE ANIMATION BY PETER FOLDES.

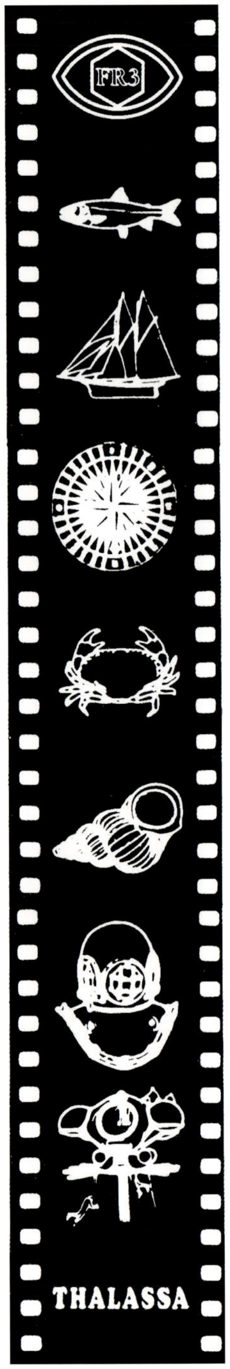

CREDITS FOR THE TELEVISION PROGRAM *THALASSA* DESIGNED BY GERARD MARINELLI.

constant increase in computing power soon introduced the possibility of a new dimension in animated digital imagery: real time. In 1982, Silicon Graphics marketed its first Iris workstations that could generate in "realistic rendering" several tens of thousands of facets in a fraction of a second[11]. The image could be inverted instantaneously in all directions and be observed as if from afar or in close-up detail. At the time, computer graphics programmers would move heaven and earth to get their hands on tools of such power.

Military industry was not far behind. During these same years, flight and tank-control simulators, which projected images filmed from models onto video screens installed in place of the cockpit windscreen, were ousted by synthesized images and computers that could factor in a vast array of parameters from fog to simulated mechanical and/or electronic failures. In 1982, Claude Méchoulam opened the doors of the Sogitec[12] together with André Martin, a researcher at the Institut National de l'Audiovisuel, who there made the first fiction film using synthesized images to be produced in Europe, *La Maison Vole*. From this firm emerged Explore, later marketed by Thomson Digital Image, as well as that indefinable "French touch," a specifically Gallic expertise in digital imagery that is still appreciated on the international scene[13].

The digital, virtual model became alive, dynamic. It evolves over time. First of all, the observer can turn around it, zoom in on it or move back from such or such a detail—but also its object characteristics change in time. This includes *every* characteristic: not only its position, but its form and texture, too. Hair grows, a material becomes transparent or, conversely, turns opaque, while environmental parameters simulate night falling or the onset of a light drizzle.

11 Spectacular imagery now abounds. For an overview, visit the Internet sites of the main 3D design packages: 3D max, CATIA, EnSight, Lightscape, Maya, Pro/ENGINEER, SolidWorks, Unigraphics.... **12** Sogitec is a business that works for the Défense Nationale on the design of information systems and on the development of flight simulators in particular. **13** Among the highpoints in the history of computer animation, one should mention *Tony de Peltrie* made in 1985 by Pierre Lachapelle at the University of Montreal. Since then, the eye-catching achievements have become too numerous to be listed here.

Then, there's military engineering. The Internet is teeming with sites exalting the wonders of modern technology in the service of defense projects. Equipped from head to toe at the Natick research center in Massachusetts, today's squad member packs an electronic punch equivalent to a 1980s aircraft carrier. Designed to integrate communication tools and vision enhancers, the special textiles of his uniform and the composites of his helmet are stuffed with sensors that regulate temperature and moisture, that filter out anything from gases and radiation to the stench of socks, all the while analyzing the environment, various bodily secretions, and the private's pulse. DuPont has devised wonders woven in Kevlar, Tyvex, and Coolmax thanks to which it will be possible to survive in the most hostile conditions sporting a bulletproof vest as soft as a cashmere sweater. Ratcheting this up a notch, camouflage, by integrating technologies of the electronic-ink type, will soon be able to change hue with the season. As we wait for military research to diversify into haute couture or ready-to-wear, it's a sure thing that fantastically well-funded research programs are spending million of dollars on creating invisible combat gear[14].

SLOT MACHINES

Techniques initially confined to the military soon make their way into the games market. Early on, Microsoft proposed, in addition to its operating, word-processing, and spreadsheet systems, a "flight simulator" that in various avatars continues to delight hundreds of thousands of gamers[15]. Making exemplary use of interactivity, video games first entered bars and pubs at the beginning of the 1970s in the shape of the now famous Pong, with the no less celebrated PackMan close on its heels. They blazed a trail for the pneumatic Lara Croft, superstar of a flourishing industry whose sales turnover was soon to exceed that of films in the movie house. In the artistic field, at the beginning of the 1990s, Catherine Ikam and Louis Fleri proposed *L'Autre* ("the Other"), an installation in which the synthesized image of a large face is projected

ITEMS OF COMBAT GEAR DESIGNED AT THE NATICK SOLDIER CENTER, MASSACHUSETTS.

14 One might begin investigating at www.natick.army. mil. Using a search engine with words such as "army, camouflage, smart materials" turns up hundreds of enthralling results. **15** Flight Simulator has become notorious since 9/11, as it was this piece of software which is supposed to have been used to train the suicide pilots who drove their planes into the World Trade Center and the Pentagon.

onto the wall of a darkened room. When the viewer enters the room, the face comes alive and responds, and a strange relationship begins.

INTERACTIVITY

Popularized by games, a primary reflex among a younger generation that segues seamlessly from feeding-bottle to console, dynamic imagery and interactivity have gradually become essential facets of design. Reactivity, interactivity, the computer's capacity to recalculate an entire scene in real time, factoring in the actions of the operator controlling the program and the various misadventures envisioned by the storyboard ... interactivity soon made its presence felt in every digital application. The behavior of a digital object depends on a destiny fixed within the program, on the user's options, and on what it perceives of the external world. The immutable object whose structural stability vouchsafes its solidity has been evinced by something that has more in common with a living organism. The entity develops, morphs, adapts. A mobile phone starts off by learning the list of phone numbers of the owner's contacts, but shortly its morphology will evolve into a structure that transforms over time and with its inhabitants' changing lifestyles, before disappearing and becoming a venerable ruin or a patch of waste ground on which a new building will rise.

The marks of time do not appear only in the guise of signs of wear and tear. The object matures, it grows, and evolves. The addition of a visual sensor to a telephone, of a GPS (global positioning system) to a video camera, or of GPRS (general packet radio service) Internet access to an mp3 player is much more integrally rooted in the structure of the entry-level object, more like a graft than a mere "bolt-on." A functional protuberance grows out of the object; it is corroded and patinaed by preprogrammed rust. The system-object evolves through its interactions with its end-user and the environment: it learns how to react from the material on which it operates.

Its senescence and death, even its posthumous destiny, are programmed in. Beyond mere obsolescence intended to sustain a market predicated on self-renewal, in a more ecological perspective it is designed to manage developments—so-called "progress"—through a "softer" approach to the depredation of resources, to the scars and blemishes it inflicts upon our old planet.

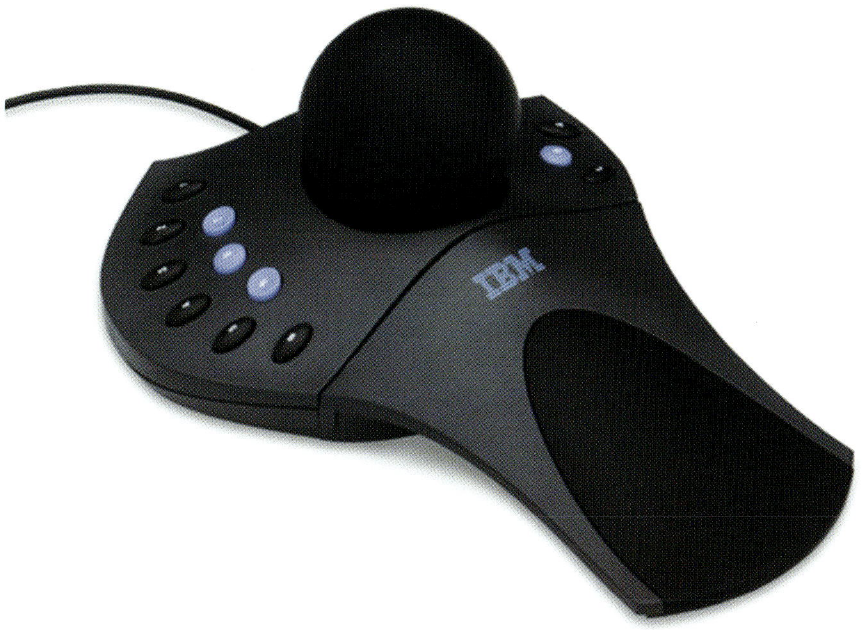

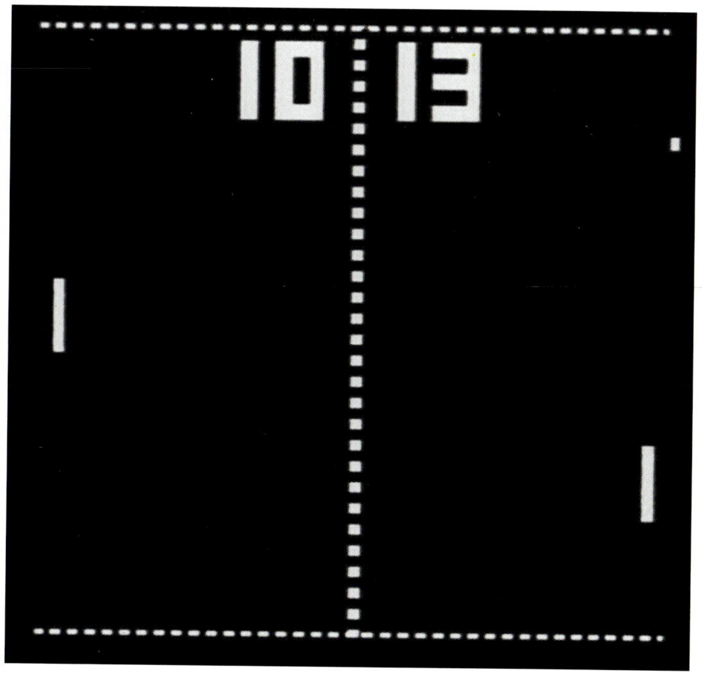

above

THE SPACEBALL, A 3D MOUSE CREATED BY LOGITECH FOR IBM.

right

PONG, THE MOTHER OF ALL VIDEO GAMES, DEVISED BY WILLIAM A. HIGINBOTHAM IN THE 1960S.

THE DESIGNER IN THE SERVICE OF COMPUTING: FIRST, THE BOX....

At an early stage, the computer itself became a design object. And its first incarnation was as the box. In 1977, at a time when Bill Gates had not yet founded Microsoft, Steve Jobs was already showing interest in design. The launch of Apple II, which Jobs conceived in collaboration with another Steve—Steve Wozniak—was presented through a marketing environment that offered an instructive contrast with the "techno" appearance of other computers of the time. In particular, great care was taken with documentation and packaging, factors that were clearly instrumental in the make's success. In 1984, the appearance of the Apple Macintosh constituted a genuine revolution in the world of computing—heralded in a less publicized manner by that of its precursor, Lisa, the previous year.

Steve Jobs' confidence in design never faltered and his return to the head of the company after being sidelined during the Sculley/Spindler/Amelio years marked the triumphant return of a design with pure lines conceived by Jonathan Ive, a design which chimed with the kind of mass-market electronics made by Braun or B&O, but which was a far cry from the rather stark, near military designs of the majority of other manufacturers[16].

...THEN SOFTWARE DESIGN, THE ART OF THE INTERFACE, AND THE ICON

But quite apart from the box, the Apple Mac ushered in another revolution—the operating system. When it came out in 1983, the Mac had no hard disk and required two 400 KB diskettes, but its creators soon unearthed two revolutionary ideas at the Palo Alto Research Center (PARC)[17]: the mouse and the graphic interface.

16 A few benchmarks: Lisa, the precursor of the Apple Macintosh, was marketed in 1983. The Mac came out in 1984 with Steve Jobs being unseated the following year. Sculley entrusted the design of the Apple IIC and its subsequent manifestations to Frog Design. The portable Mac

Unlike with other products, its little screen displayed documents as black text on a white ground exactly as the "image writer" would print them and, still more crucially, the commands were controlled with the mouse and the clickable icons, which, from this point on, became indispensable.

To general rejoicing in the publishing and graphic design trades—which adopted it all the more readily as Adobe marketed some splendid dedicated graphic design software—the arts of typography and of page-setting could be reinvented; it all recalled the Ulm school and the typographical inventiveness of the 1920s. Graphic interfaces, with their erogenous zones titillated by the mouse, became the IT standard.

Employed initially for his competence in graphics, the designer soon gained weightier responsibilities with software design. Starting out from various functions and utilities, his job was to define the interface, the ergonomics, and even the functionalities of the programs or the scripts kicked off by the interfaces. And all this occurred at the precise moment when computer programming was undergoing profound upheaval. Not only was the "program object" making it possible to play as if with bricks that could be assembled and described in terms of functional characteristics and their capacity to be coupled to other objects, but the "top-down" approach, which starts out with global functionalities and only then descends to fundamental elements (the raw material of the IT specialist) became well-nigh universal in programming.

Interface design was no longer an outcome of programming or the final lick of paint—it had become the starting-point for the whole project. And in the process, apart from the interface, the designer's creativity extended to the whole script that unfolds from the time one inserts a coin in the slot and selects "weak or strong, with or without sugar." Inhabited by computer, the object has become dynamic.

launched in 1989 was a commercial flop. It was also a design failure. Apple was in freefall until 1997, when Steve Jobs took the helm again and design returned to the forefront of the firm's concerns, in particular in the products designed by Jonathan Ive's team: Imac, G3, G4, G5, PowerBook, Ipod....

17 Palo Alto Research Center. Xerox's research center.

above, left

THE FIRST CORDLESS TELEPHONE:

DYNA TAC 8000 BY MOTOROLA,

1983.

above, right

THE NOKIA 8110 MOBILE

TELEPHONE, 1998.

DIGITAL MATERIAL

Imperceptibly, the designer has been pulled into the fray. As he drew the first icon, he signed a pact. He became part of a brotherhood not without mystical connotations. The result is dematerialization: earthly bodies vaporize into virtual reality even in cybersex, promising the possibility of not only simultaneous presence at every point on the planet, but even eternal life in the form of binary memories that, so we are told, are indestructible.

Incontestably, the Internet marks a major revolution in the history of humanity. If humanity can avoid the ecological and military apocalypses it is so busy preparing, the technical, economic, and cultural repercussions to come will probably exceed our wildest dreams.

Once again, the most profound transformation will stem from a new way of looking at human products. To the concepts of dynamism, interactivity, and evolution can now be added the concept of collaboration; the "one plus one equals three," or, the whole is far more powerful than the sum of its parts.

The global network, the Internet, the mother of all networks, interconnects components that are themselves networks. The interbank network, through which money circulates that is no longer material, connects together the companies' own facilities, themselves composed of hundreds, if not thousands of computers. At desk level, the computer, one element in a much vaster service, can itself be thought of as a tiny network in which information is exchanged between its components—keyboard, mouse, screen, and disks.

Not unlike a set of fractal Russian dolls, in which each element reflects the complexity of the whole, what is true on the largest scale is also true at the microscopic level. Just as for object-oriented programming, the physical infrastructure is described in terms of building blocks with characteristics and capacities conceived to achieve a given function and to communicate with others. The computer senses the presence of a connection point to the Internet over radio waves. The camera is detected by the telephone, which has also pinpointed a desktop speaker located a foot or so away. Once mutually aware of their presence, all these elements start to communicate with each other. They understand one another because they all respect the

following pages
POWERGLASS: A COMBINATION OF A SERIES OF INCREASINGLY SMALL, FLAT AND POWERFUL LEDS (LIGHT-EMITTING DIODES) AND THE TECHNIQUE OF LAMINATED GLASS. THE DIODES ARE FUSED BETWEEN TWO SHEETS OF GLASS. THE POWER SUPPLY IS COMPLETELY INVISIBLE AND ALLOWS EACH POINT OF LIGHT TO BE FIRED INDIVIDUALLY, PROVIDING THE MATERIAL WITH A VAST RANGE OF INFORMATION AND DECORATIVE POSSIBILITIES. MANUFACTURED BY KNELL PLATZ, GERMANY.

same protocol. In the Bluetooth protocol, for example, one of the first things they do is to decide who, in this new relationship, will be the "chair," who will arbitrate their decisions. The new helpmates then pool whatever they can bring to the community and search within their memories what they can achieve in partnership. On occasion, the "chair" will consult the Internet to discover if anyone has imagined an unexpected use for their collective muscle. Once the control program is booted up, the interface of this new system is reconfigured and additional functions are proposed to the user.

The mobile telephone serves as a master key to the vast network that connects all the inhabitants of the globe and is in itself a mini network with ever more extensive functions. It evolves so speedily that users sometimes, understandably, resent that a product costing a good thousand dollars today is obsolete six months later. The most recent model doesn't just have an integrated digital video camera, a GPS receiver, and all the functions of a digital organizer; it is also delivered with twenty hours of free music to download from the iTune's catalog.

And tomorrow? If every appliance with even a limited amount of electronics has an address and an identity, as will soon be the case with the arrival of the IPV6 standard, then the most seemingly off-the-wall dreams of the automated home will outstrip the wildest science-fiction scenarios. The presence of a particular member of the family at a particular hour in a particular room will influence the light, sound, and even olfactory environments, which will be scheduled in accordance with habits learnt by the home-management computer. The telephone will interact with intruder detectors, and temperature sensors regulate the air-conditioning[18]. A multitude of electroluminescent diodes and flat screens will connect the lounge, kitchen, bedroom, and office with the universe, in all its dimensions, from the infinitely large to the infinitely small, real or imaginary, all dissolved into the same universal binary representation. Wireless transmission technologies, such as Airport, Bluetooth, and the

18 In *The Martian Chronicles* published in 1950, Ray Bradbury describes a house that controls its inhabitants' morning alarm, their musical and light environment, their work agendas and breakfast. As the field widens, however, the house is seen to be the last construction still standing following a nuclear cataclysm that has annihilated all traces of life from the surface of the globe.

rest, are gradually eliminating the unsightly tangles of wire peeking out from under the TV, stereo, and computer.

Thus the 1950s food processor will have a second wind. The vacuum cleaner will be alerted to every nook and cranny in the house and have the tact to come automatically out of the broom cupboard only when the house is empty. Naturally, it will be able to sort through whatever it vacuums up ... and if it finds a coin or a treasured ring beneath the chest of drawers, it will simply put it on a side-table. In the utility room, the recently launched Stronzetta's advanced system has it covered: it will analyze the family laundry to determine the exact temperature, the right mix of detergents, and the correct quantity of water required for a perfect wash and to maximize respect for the environment.

Saint-Gobain, a French supplier of building materials, now carries a type of glass that could only be dreamt of ten years ago. Thanks to electrochromic windows, shutters are no longer required to plunge a room into complete darkness: a burst of electricity will turn the pane opaque. This type of glass also features in interiors to demarcate the space and ensure a measure of privacy. On the other hand, as night draws in, light will continue to pour from windows equipped with electroluminescent polymers. Analogous materials can filter light to create an antiseptic environment suited to bringing plants to fruition or to medicalized or other extreme environments, such as subaquatic or outer-space habitats.

following pages

INTEGRATED CIRCUIT, A TEXTILE THAT THINKS IT'S AN ELECTRIC CIRCUIT. INCORPORATING CONDUCTING WIRE IN ITS WEAVE, IT CAN BE MADE INTO ANY TYPE OF FUNCTIONAL ELECTRIC CIRCUIT. THIS TECHNIQUE HAS ALREADY MADE IT POSSIBLE TO WEAVE VIABLE BARCODES, AFFORDING THE PROSPECT OF COMMUNICATION-READY CLOTHING FITTED WITH DIODES AND ELECTRONIC EQUIPMENT. MANUFACTURED BY AVERY DENNISON RINKE, GERMANY.

Little by little, though, glass is retreating before polycarbonates (which are much lighter, much more resistant, but unfortunately as yet much more expensive) in which it is comparatively simple to integrate various layers of electrochromic smart films that in the future will become as adaptable as any screen.

A bit of programming and the mirror above the mantelpiece will become a video screen. Not only in interior installation, but also in telephones, micro- and nano-computers, electronic organizers, cameras, video recorders, GPS ... new applications for electroluminescent polymers (OLEDs[19]) are being announced every day in all fields.

In the wake of photochromic glass whose opacity varies with ambient luminosity, the eyeglass industry is conducting research along similar lines.

AL CIRCUIT

Made lighter and more compatible with a city suit, the Natick soldier's "augmented reality devices" will surely make an appearance on the open market one of these days and become the essential accessory to a communication-ready wind-breaker and a pair of laterigrade sneakers.

The future might bring a digital car with GPS, SatNav, and telephones; its Sekurit electrochromic windows and windscreen, its anti-collision detectors and other safety devices will no longer need miles of electric cable thanks to the digital mini-network on carrier currents[20], but it will nevertheless require a new 48-volt battery to supply all its peripherals.

With one eye on medical research and the other on advances in biochemistry, some have even begun to dream of symbiotic man[21]. Since the beginning of the 1980s, neurons have been grown on silicon substrates. The objective is clear: to restore mobility, sight, etc., to wounded or invalid patients by improving the interface mechanisms able to compensate for a defective body. The ethical debate is an open and shut one when it is a question of just repairing; but once the borderline shifts imperceptibly towards eugenics the debate is less clear cut, and the idea of producing a new master race awakens bad memories.

Situations quite unimaginable today could become common currency for our great-grandchildren. Fiction is packed with these hybrids, half-man, half-machine, which either strive to defend order and morality in a world torn apart by violence or which turn an almost ideal society, where the notion of work is delegated to (over)zealous robots, upside-down. Every possible scenario has been explored and it is difficult to settle on any particular one—especially since it is enough to look at the debates over abortion or euthanasia to see that, as the years pass, bioethics is sure to evolve.

19 With an energy efficiency considerably higher than traditional filament bulbs and easy to produce in microscopic sizes, OLEDs (Organic Light Emitting Diodes) have ushered in a genuine revolution in the world of lighting. **20** By substantially reducing the complexity of the electrical profile, transmission currents allow digital information to circulate along the electrical supply network and be recognized by recipients as being addressed to them. Multiplexing in this manner obviates the necessity for a cable for each appliance. **21** For more on this topic, see any one of the numerous publications by Joël de Rosnay.

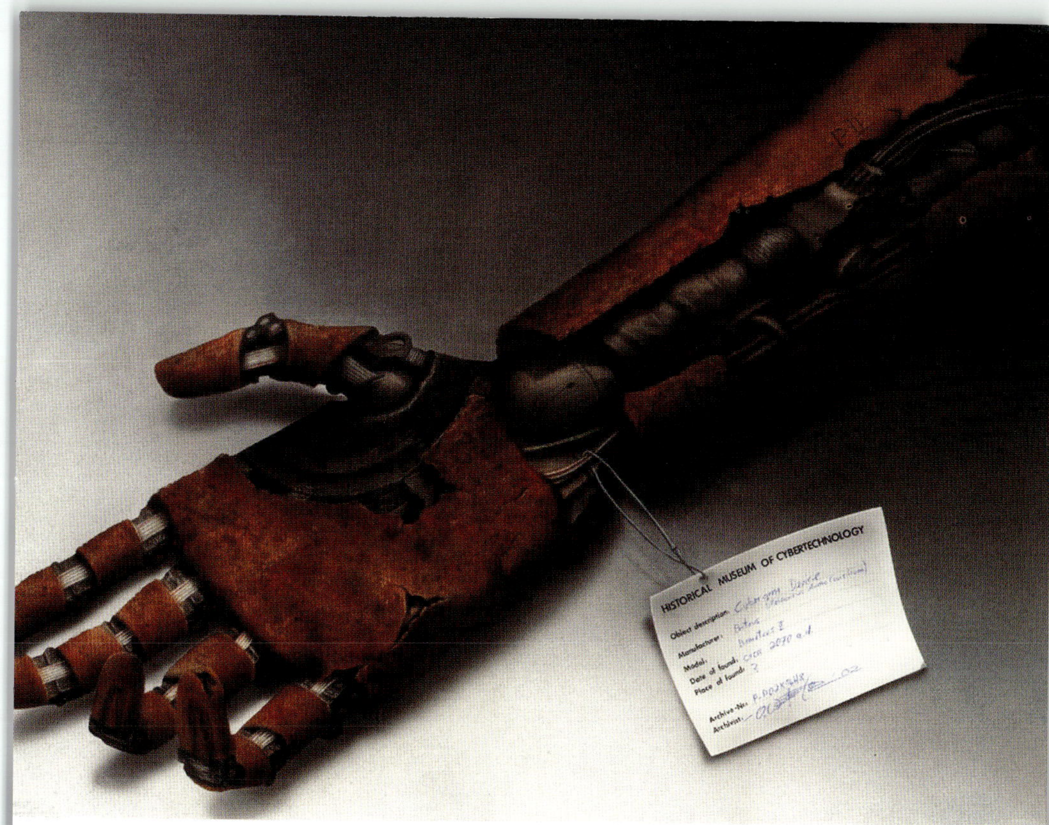

WITH *PROTEUS II*, THE ARTIST
OLIVIER WETTER PROPELS US
INTO A FUTURE WHEN ANDROIDS
WILL BE MUSEUM PIECES.

ASIMO
HONDA

The intersection between biochemistry and digital technologies is an extremely fertile field of research. Putting aside for the moment the fundamental question of consciousness (not moral conscience, but cognitive consciousness) and, in a more general way, our customary definitions of life—be they religious, philosophical, or mechanist—the analogies to be made between digital programming and biogenetic programming are self-evident.

Biochemical calculators are already being tabled and are one of the most promising paths in future IT development, on a par with quantum computing. One dreams of digital ribosomes inhabiting programmable materials, of sensors and actuators of every kind, in a marriage of biology and electronics under the rolling eyes of a Doctor Frankenstein and his new dog, Aibo.

Nanotechnology was supposedly born at a lecture given by Richard Feynman in 1959 entitled, "There's Plenty of Room at the Bottom."[22] Richard Feynman, Nobel laureate for physics in 1965, was probably one of the most productive physicists of the postwar period. During his lecture he explained why there is no theoretical barrier to manipulating matter on an atomic scale. Specifically, he showed how it might be possible to miniaturize the twenty-four volumes of the *Encyclopedia Britannica* to the size of a pinhead. And this was in 1959. Subsequently, it was probably Eric Drexler who worked the most on the idea, dreaming up scenarios that were to inspire science fiction authors. Yet these ideas did not remain on the drawing-board—far from it.

In 1970, Gordon Moore, one of the founders of Intel, the world's largest microprocessor manufacturer, stated the law that bears his name: the number of transistors in an integrated circuit per unit volume, on which computing power and memory size directly depends, doubles every eighteen months as its cost halves. How far will it go? Certain specialists predict that this descent into the infinitesimal will come to a halt in 2010; others reckon there is no reason it should stop before 2020, by which time the potential for those technologies destined to take over from the microprocessor—quantum and biological computing—would have been thoroughly explored. The hundreds of

facing page
FOLLOWING DOCTOR FRANKENSTEIN'S CREATURE AND VAUCANSON AUTOMATS, HONDA'S ASIMO ROBOT IS, WITH SONY'S DOG AIBO, A MILESTONE IN THE PURSUIT OF THE AGE-OLD DREAM OF THE GOLEM.

22 This premonitory text can be found on the Web, for example on the Xerox site (www.zyvex.com/nanotech/feynman.html)

thousands of transistors making up a microprocessor measure only 50 nanometers. The basic logic gate measures 15 nanometers[23].

Admittedly, these are only electronic circuits. But, coupling such circuits with electromechanical polymers that can already be manipulated on a microscopic scale makes it possible to imagine nano-machines with the dimensions of a large molecule.

Currently, physics does not countenance reducing a human being to the size of a red blood corpuscle, as in Richard Fleischer's 1966 movie, *The Fantastic Voyage*. But today, nano-cameras the size of a multivitamin capsule can film inside the digestive system, and whole laboratories on microchips of comparable size can analyze drug molecules in real time and rearrange them in response to the findings. Molecular machines are not just a dream and will one day be part of our children's environment. Powered by electronics, intelligent substances acquire a new facet. Thus, in 1993, John Storrs Hall proposed his concept of Utility Fog. These are molecules a few dozen microns in size that are capable of communication and self-assembly. Together they form a parallel calculator of massive power and can adopt any form desired. Better still, in drawing closer together or farther apart, they vary the density of the object formed, while the structure of the molecular network determines the color, not unlike the iridescence on a butterfly wing.

With this type of material we've come full circle. As long as they can be described in a program, more or less complex objects can be readily manipulated. After all, the production chain that makes Jacobsen's Ant chair is surely less complex than for a daisy or a lily-of-the-valley. Perhaps one day our children will farm chair trees? The aquatic protuberance on the extraterrestrials in *Abyss* or *T1000* art directed by James Cameron emerged from brains like those of IT experts.

Working on the infinitesimal scale is no longer a job for the engineer or the atomic scientist alone. This future is part of the designer's dream, too;

23 To appreciate the scales we are talking about requires a little time. A kilometer is clear enough, as is a millimeter, and one can grasp how the one compares to the other. The technologies referred to transform a millimeter into a kilometer by photographic reduction. At this level, 50 nanometers correspond to 5 centimeters. A water molecule measures approximately 2 Angström units across, on this scale, 2/10 of a millimeter.

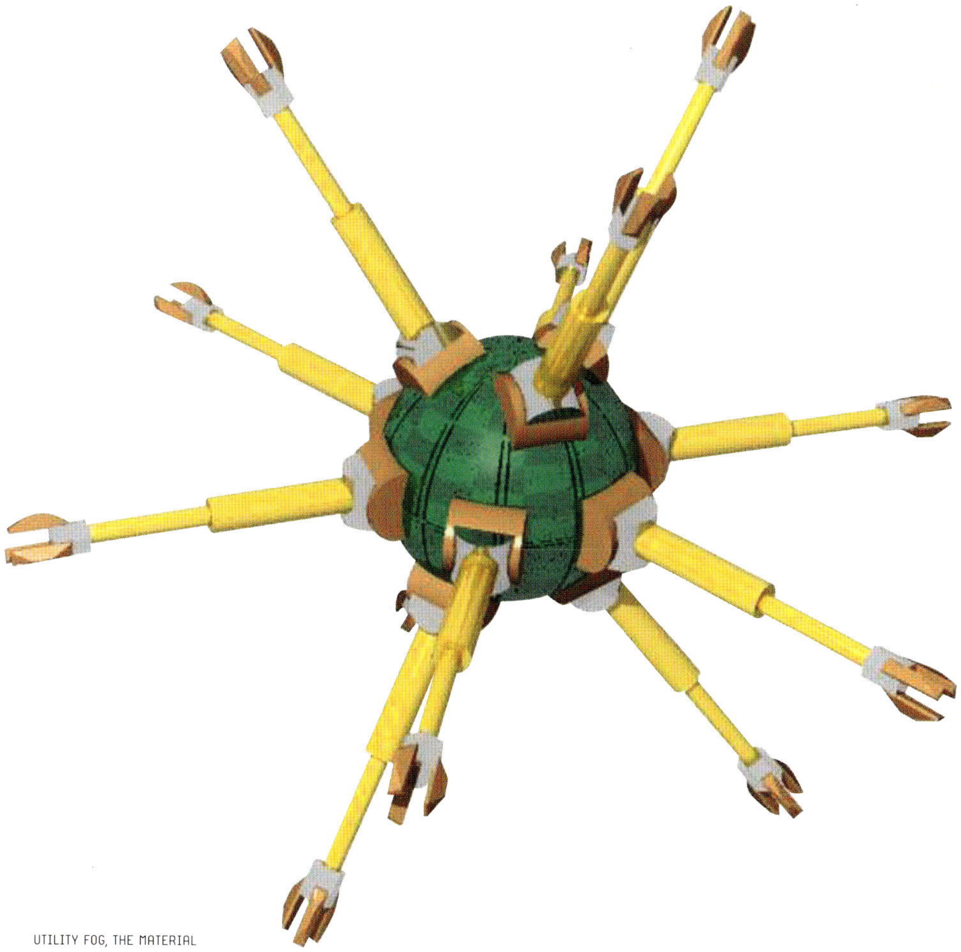

UTILITY FOG, THE MATERIAL
DREAMS ARE MADE OF.
DEVISED BY JOHN STORRS HALL,
UTILITY FOG IS FORMED OF TINY
INDEPENDENT CELLS OCCUPYING
A SPACE ONE-HUNDRED MICRONS
ACROSS. THE CALCULATING AND
COMMUNICATING CAPACITY OF
THESE CELLS CAN BUILD INTO
A MASSIVELY PARALLEL COMPUTER
OF HUGE POWER, WHILE ITS
TWELVE JOINTED ARMS ENABLE
IT TO ADOPT ALL CONCEIVABLE
FORMS AND DENSITIES.

one which can be shared in the form of scenarios substantially more realistic than the books of science fiction that sometimes show scant regard for science and technology[24].

Strenuous efforts are required to imagine the realistic applications and consequences of working on scales like these and they call for a critical assessment of such sci-fi storylines, as well as continually monitoring technological advances. Fortunately, the latter task has been made less indigestible through the abundance of worthwhile sites on the Internet dedicated to the relevant techniques[25].

SILICON, A NEW RAW MATERIAL?

It's all enough to make your head spin. If designers want to assert their right to dream of the future, if they wish to advance an ethical position through their work and to make a worthwhile contribution to shaping the world, very soon they will be confronted by unavoidable questions. Can one truly believe that IT generates social equality, and is the socioeconomic world it generates encoded in a manner the designer can accept?

On the one hand, some will refuse to have anything to do with it: leave it to politics, philosophy, or religion to cordon off the no-go areas. But with ever-advancing technologies, black-and-white arguments about the sanctity of life become ever more difficult to uphold. How, today, do we define life? What is consciousness? We can be pretty sure that these questions will remain unanswered for a long time yet.

This incertitude can breed another attitude: one which confidently soars onwards and upwards, promoting further explorations, without ever pausing to think that knowledge spawns civilization, and ignorance, barbarism. In a time when laboratories come up with ways to make the most outlandish

24 Neal Stephenson is one of the writers of science fiction who best describes nanotechnologies. In *The Diamond Age*, a host of such applications form a facet of the rather somber world to come. It is interesting to note that the majority of storylines featuring nanotechnology present a very pessimistic view of an inhospitable and violent world. **25** A good starting-point is the Foresight Institute site (www.foresight.org)

ideas possible—or even outstrip them—the optimists are having a field day. Meanwhile, pessimists are tearing their hair out at the idea of standing by as life disappears from a soon-to-be vitrified planet. The overwhelming majority sails along from one compromise to the next.

A blasé, cynical designer, who just responds to the demands of a marketing strategy that describes the market in terms of social styling, will dream up products intended for adepts of cool cocooning who, with a spliff between their lips, lounge about under digital duvets to the sound of soothing music from a DVD on a flat screen and in a light environment wholly programmed by electroluminescent diodes. Closely related is the nomadic, cocaine-snorting jet-setter, whose every step is monitored by an eagle-eyed GPS satellite and who lugs about a stack of digital gizmos, all offshoots from military research. The chip is a raw material like any another: Asia's brainchild has been adapted and formatted to milk the market.

A two-speed society? A two-speed digital world? Towards the end of the twentieth century, political voices were raised to propose a full-frontal assault on such inequalities in an effort to bridge the social abyss. Unfortunately, it's already clear that many elected officials, those in the know and those as yet unseeing (all the usual victims) will probably not be invited to the great junket. Today nobody dares assert that IT will make social inequalities disappear.

Though the backdrop may be distressing, it is no time for designers to fall asleep on the job. Overall, their value system tends to encourage listening to other people and an open attitude to human relationships. They have to observe and to listen to others with an open mind, so as to understand, consider, select, and, only then, make proposals. The best course of action for today's designers probably consists in offering scenarios that incorporate all their ideas, and communicating them in such a manner that everyone concerned can make their choice with the clearest possible understanding of the likely consequences.

Design research is nourished by an acute technological curiosity at least as much as by an omnivorous appetite for culture, both interwoven with a personal morality whose validity needs to be tested at any opportunity. To do this one has to devise plans with every means possible, to present them and share them with the largest possible number of people, with the aim of ascertaining their strong and weak points, their potential and their pitfalls.

ECOLOGY SELLS

JEAN-JACQUES SALOMON

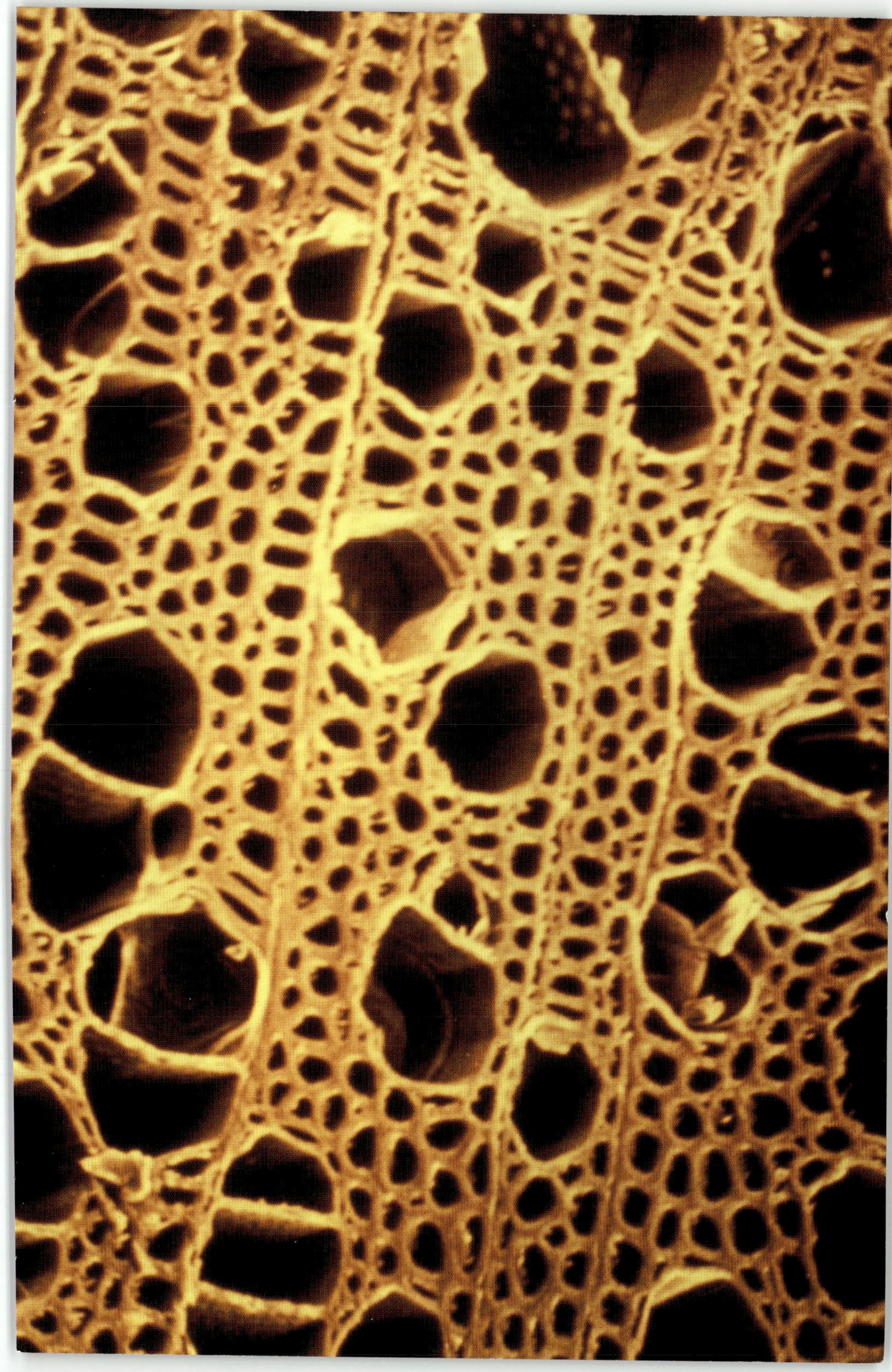

Raymond Loewy's book *Never Leave Well Enough Alone* was compulsory reading for exponents of the new industrial design of which Loewy was one of the founding fathers in the United States and one of the pioneering forerunners in France. The book arrived at just the right moment for a Western Europe that was slowly emerging from the cataclysm of war. In spite of shortages and even hardship, the period of reconstruction in the strict sense was over; the spurs to production set up by the Marshall Plan had initiated European industrialists and trade unionists alike to both the concept and the techniques of American management. In a way, Loewy's book served to point out a hitherto unknown aspect of the ingredients essential to the success of a technical innovation: the association of marketing with art. France, particularly, was entering a period aptly dubbed "the thirty glorious years," an era of impressive growth and full employment and of a consumer and soon leisure society blessed with a plethora of domestic durables. Surprising and provocative in equal measure, Loewy's text acted both as a signal for and an illustration of this period.

What Loewy wrote on his arrival in the United States shortly after World War I of his revulsion for mass production, admittedly highly efficient, but cursed by coarse and ugly objects—from tools to furniture and machines—"without the rational economy of means or materials," could equally apply to Europe shortly after World War II. America was entering a period of unheralded industrial growth. As Loewy wrote:

It was a sellers' market and an era of unprecedented prosperity. The entire production was absorbed by an eager public fixated on performance alone and in spite of appearances. The market flooded with refrigerators perched on spindly legs, and others topped by clumsy towering tanks. Typewriters were enormous and sinister-looking. Carpet sweepers when stored away took the greater part of a closet, and telephones looked (this is no pun) disconnected. I felt that the smart manufacturer who would build a well-designed product at a competitive price would have a clear advantage over the rest of the field when things became tough....

preceding page
DETAIL FROM A CROSS-SECTION OF A PIECE OF POPLAR.

facing page
RAYMOND LOEWY IN THE EXHIBITION SPACE FOR THE *DESIGNER'S OFFICE AND STUDIO*, METROPOLITAN MUSEUM OF ART, NEW YORK, 1934.

following pages
PREPARATIONS FOR THE SALON DES ARTS MÉNAGERS AT THE GRAND PALAIS, PARIS, FEBRUARY 24, 1954.

L'ELECTRICITE
EST ECONOMIQUE
NORD
OCEL

Crio
STIC B
ABEILLE
CIRES
SULTANE
A BRULER
SORTIE
VISITEZ LE SALON LE MATIN
SORTIE
Lemercier Frères
SORTIE SORTIE
IS - RHÔNE
ASPIRON
SAUTER
CHAUFFE-EAU ELECTRIQUE Lemercier Frères
SAUTER

THE ERA OF INDUSTRIAL DESIGN

The words "harmoniously," and "harmony" recur throughout Loewy's book And he never stopped reiterating this principle, as if he had learnt one of the current definitions of the word by heart: "A group of characteristics (a combination of sounds, accents, rhythm…) which render a speech pleasant to the ear" (Boileau), or else, "a rapport, a relationship between parts of a whole that makes these parts combine to the same overall purpose" (Richelet)—in short, for Loewy, harmony is the relationship between form, material, and function that contributes not only to an article's attractiveness but more particularly, to its market success.

By the time Loewy's book appeared in 1953, France had only just emerged from its wartime economy and was not yet ready to think of form playing a role as—if not more—significant as material and function in industrial product design. Nevertheless, there were glimpses of the future triumph of the consumer society, with its abundance of commodities, old and new, defined or refined by specifications as much as by aesthetic appearance, some useful and others completely useless. Just as France was about to enter its thirty-year period of prosperity, few were pondering the tricks of the market, the pointlessness of sales systems dominated by advertising, or the limits to, still less the harm inflicted by, progress. Quite the contrary: consumer goods from American society looked like marvels that had materialized miraculously in the "Yankee" Ali's Baba cave; they were products dripping with a happy-go-lucky aesthetic, at once motor and guarantor of an irresistible growth curve in sales.

It was time to march resolutely toward an era of rationalized forms that, by improving outward appearance and function together, conspire to convince customers that the more they consume (from toothpicks to automobiles, from cigarette packets—ah! the passage from green to white on that symbolic pack of Lucky Strike!—to bus conductors' uniforms, from refrigerators to locomotives), the more likely they are of acquiring eternal bliss. For sure, America was not without its own critics of a consumer society whose flummery and abuses—"subliminal advertising"—hiding beneath the smooth exterior of marketing and publicity campaigns Lewis Mumford and Vance Packard had already denounced. But for now France was still, as Marcel Bleustein-Strainer has noted, at the hard-sell stage and *L'Observateur* newspaper, at the time free

ADVERTISEMENT FOR LUCKY STRIKE CIGARETTES WITH THE PACK DESIGNED BY RAYMOND LOEWY, *THE SATURDAY EVENING POST*, UNITED STATES, AUGUST 1951.

THE SATURDAY EVENING POST
When you've been in and had a swim,
A cigarette tastes great.
But Luckies taste the best of all,
So, friend — why hesitate?
ies taste better than any other cigarette!)
Be Happy-Go Lucky!
LUCKIES TASTE BETTER THAN ANY OTHER CIGARETTE!
Fine tobacco—and only fine tobacco—can give you the enjoyment of a better-tasting cigarette. And L.S./M.F.T.—Lucky Strike *means* fine tobacco. That's why you'll find that Luckies taste better than any other cigarette. So, Be Happy—Go Lucky! Get a carton today.
At Anacostia Station, folks,
There is a favorite brand —
It's Lucky Strike that we all like,
The smoke that sure tastes grand!
(Luckies taste better than any other cigarette!)
A LUCKY STRIKE FOR ALL AMERICA!
BUY U.S. DEFENSE BONDS
LUCKY STRIKE
"IT'S TOASTED"
CIGARETTES
L.S./M.F.T.
COPR., THE AMERICAN TOBACCO COMPANY
S TASTE BETTER THAN ANY OTHER CIGARETTE because...
/M.F.T.- Lucky Strike Means Fine Tobacco

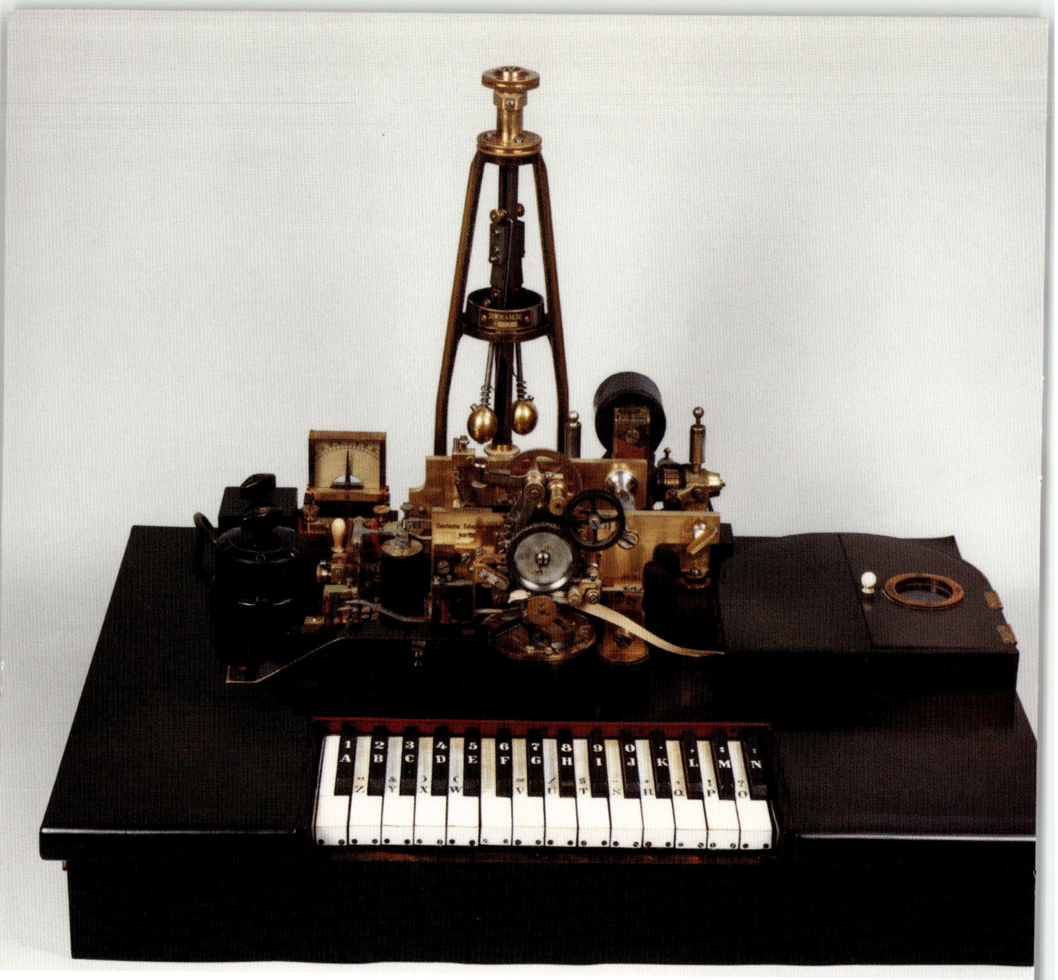

DAVID EDWARD HUGHES,
TELEGRAPH WITH ALPHABETICAL
DIALING APPARATUS, 1880,
POSTMUSEUM, BERLIN.

from advertising as if through monastic vocation, was not yet on the point of heralding "the generation of Marx and Coca-Cola" or diving into the delights of unbridled consumerism.

Of course art had always been inherent in the manufacture of tools, from prehistory onward. In a later era, the clock-making industry, for instance, or the manufacture of oceanic navigation equipment and scientific instruments had not waited for the twentieth century before making use of "design before design," that is to say, before adding beauty of form and material to out-and-out function. But it remains true nonetheless that it was mass production that turned design into one full-blown industry product among others.

Incontestably, Loewy was the living embodiment of the advent of a new professionalization (he was the founder in the United States of the earliest companies specializing in industrial design), and what he said of America after World War I undoubtedly applied to Europe after World War II: "So here we were: saturation was at the door; competition would become fierce, good design could help sales, manufacturers could be convinced—and I was the one to do the job."

In the boom decades, the Angel Gabriel of design saw his annunciation confirmed in Europe beyond his wildest hopes. Mass consumption went hand in hand with the expansion of a design profession dedicated to enhancing "beauty through function and simplification," as he put it. Soon designers found themselves at the forefront of every stage in the production process and companies put them on the payroll, their opinions and models having a greater impact (together with those of marketing professionals) on the decisions taken by CEOs or boards of directors than those of the white coats in the labs. In France it was the time when prestigious design schools and diplomas in industrial aesthetics came into their own: the institutionalization of design has made it a profession to be reckoned with.

A TIME OF QUESTIONING

Since then though, the economic and social landscape has altered beyond recognition. For more than half a century now we have been in a global situation, with an economy and means of production and consumption that are very different from that which governed the years of plenty. The changes of all kinds that affected the close of the century—from the fall of Communism to seeing airliners transformed into torpedoes crashing into the towers of the World Trade Center—have brought a measure of uncertainty into international relations, which calls into question many generally accepted ideas on the equilibrium that, in spite of everything, the stand-off of "mutually assured destruction" had maintained. Globalization has so transformed the world economic situation that we are witnessing certain industries from the north relocating factories and services to the south, while some firms from the south are buying their counterparts in the north.

Already at the end of the 1960s, the Meadows Report, "Limits to Growth," had sounded the alarm as to the possibility of indefinite growth in industrialized nations. The oil crisis of 1974, catalyzed by the war in the Middle East, and the resultant slump spawned a significant and entrenched rise in the unemployment rate and quickly saw this report shelved in the already oversubscribed category of future forecasts seemingly without a future. Nevertheless the perception of inherent limits was spreading, while at the same time—as had earlier been the case with electricity—a revolution in electronics and computing, from laptops to Internet networks, was constantly transforming and renewing every sector of production and consumption.

The debate launched by the Meadows Report resurfaces today in fears stemming from not only the exhaustion of natural resources but also global warming. Few things articulate the economic changes and shifting mindsets better than the catalog accompanying a 1996 retrospective of Gaetano Pesce's oeuvre at the Pompidou Center: "Our era is, it seems to me, no longer one of exclamation marks but rather one of question marks. In other words, we no longer live in a time of answers but in one of questions." Pesce reflected on the role of designers in a society gripped by uncertainty and proposed to refloat it by following the path of the "poorly made" while simultaneously turning to new

GAETANO PESCE, PROJECT FOR A SHOW OF HIS OWN WORK *GAETANO PESCE, LE TEMPS EN QUESTION* AT THE POMPIDOU CENTER, PARIS, 1996.

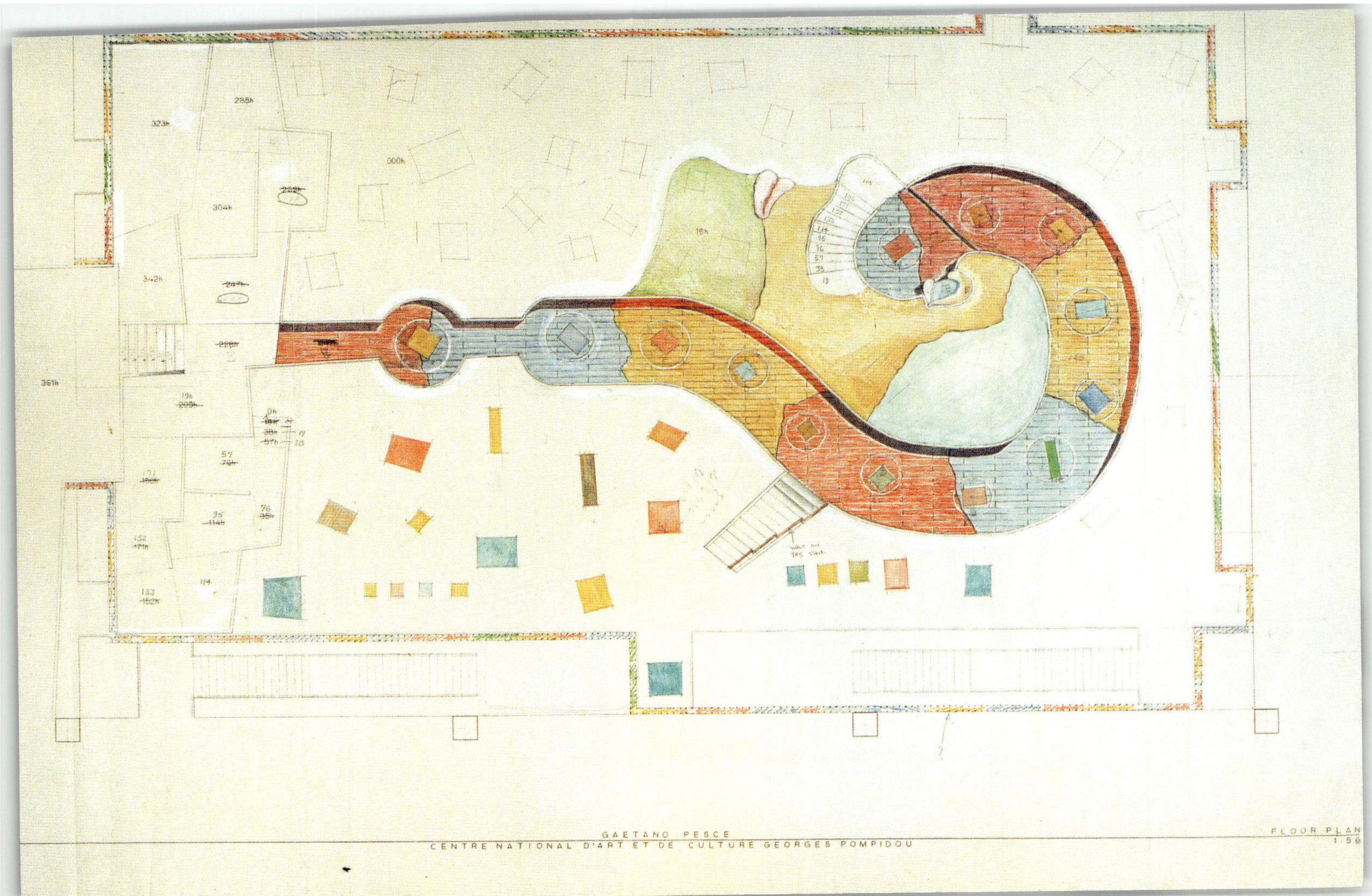
GAETANO PESCE
CENTRE NATIONAL D'ART ET DE CULTURE GEORGES POMPIDOU
FLOOR PLAN

materials as a source for "linguistic renewal." In this quest for a new order or a new equilibrium, however, designers are no less stymied than the rest of society by the ongoing upheavals of every kind—economic, political, cultural, and strategic spheres—that accompanied the closure of the shadow theater of totalitarian ideology that had beleaguered the twentieth century with its crimes and blunders.

One of the symbols of the consumer society most criticized by these new approaches is the industrialized object conceived, modeled, and fine-tuned by designers to seduce by its external appearance rather than by its function, and which is not made to last: the disposal, throwaway object.

The current vogue for new-style espresso machines, half toy, half objet d'art—magnificently designed, admittedly, with lines reminiscent of a steam locomotive—offers a dual illustration of how consumers, relieved of effort of any kind, from cranking the hand-mill to calculating the amount of ground coffee to put in the filter, have been alienated. Prisoners of a captive market, they can only purchase coffee exclusively dedicated to this machine, a product that functions solely with one given type of pack; and then, in a society increasingly uneasy about the hidden ecological costs of waste, they throw the empties lined with recyclable aluminum into the trash can. (The espresso, it is true, can be flavorsome.)

Little thought has been exercised on the propensity of industrialized societies to manufacture objects whose aesthetic qualities do not preclude them from being thrown away, (to name but a few, disposable lighters and watches, plastic or paper bags, even when they can be recycled, etc.) objects than that can never become "loci of memory," and which, unlike an archeological find that reveals a particular stage in the eternal strivings of human history, will leave no identifiable trace of civilization or culture. "The destiny of a work of art," Octavio Paz has said, "is the chilly eternity of the Museum; the destiny of an industrial object is the dump." But what of a civilization that destines a work of art to the dump? The disposable object stands outside history; it betrays a functionality that has neither past nor future, and whatever the aesthetic allure the designer has bestowed on it, it is consigned to an imaginary museum that itself can function only as a tomb. In the same way, in architecture, our civilization is the first in which *reuse* is unknown—or shunned: once demolished, a concrete, steel, and glass

REINFORCED CONCRETE BUILDING IN THE COURSE OF DEMOLITION.

skyscraper has no use whatsoever; it is crushed and added to concrete ballast for highways.

This was seen all too clearly during the demolition of the old slaughterhouses at La Villette, in the northeast of Paris, where nothing less than a fleet of magnetic mega-machines were required to extract the metals that had once reinforced the concrete blocks. The massive clumps of false stone were fated to disappear beneath the macadam of some new thoroughfare, as if they had to be hidden out of shame or bad conscience. This is a far cry from ancient cities, from the most historically remote to ultimate testimonies of the nineteenth century, whose architecture and specific aesthetic characteristics rest on successive layers of various civilizations and cultures, inscribed in reclaimed columns, marble, stone, and bricks. In the cemeteries of waste, the disposable eclipses beauty. It is quite a different relationship to time: reuse preserved memory; recycling dissolves it.

THE PARADIGM OF NEW MATERIALS

Environmental concerns—pollution and the waste piling up on the surface of the land or buried in the ground, the accelerating disappearance of animal and plant species, the inexorable exhaustion of nonrenewable sources of energy, and, more especially, the increased evidence of global climate change due to human activity—offer glaring illustrations of the price now being paid for the indefinite growth of the industrial base. Even if the world remains far from agreeing on the means to change, it is clear that prevailing models of production and consumption call for major modification, or even for a genuine conversion to a type of economic development that takes serious account of *duration*—that is, a new balance between the exploitation of natural resources, the demands of mankind, and demographic pressure.

Faced with a paradigm shift in the history of industrialization, design and its practitioners lie at the heart of the debate and, with or without the pinpricks of conscience, they are nothing short of summoned to make clear where their loyalties lie: for some, they, just like advertising agencies, are merely scapegoats in the case against industrial capitalism—fellow travelers or, worse, "sleepers" for pollution-belching companies fixated on profit—while for others, by devising new objects produced and consumed in accordance with the ambitions and the virtues of an economy that is compelled to become increasingly "green," they constitute a privileged bridge between a company and its customers. In fact, some designers are already working for a future of more viable (not "durable" for reasons that will be clear below) development, making use of both new materials and new functions, just as there are firms (and consortiums) who are singing from the same hymn sheet and claim that they now accord equal priority to techniques mindful of the environment.

Hence design has to take into account the end of an item's useful life: products, or in any case their components, need to be made as recyclable as possible. In all countries, regulations are tending to make the manufacturers responsible for their output throughout its lifecycle, a constraint that encourages reuse, reconditioning, or recycling at every stage of an article's existence even once it leaves the market structure. The system is not only operational, it is profitable too, as efforts in this direction carried out by Xerox, Electrolux, Kodak, and Fuji have shown: it is not for nothing that ecological value now forms an integral part of marketing campaigns. Even as it was being set up, the French agency for the promotion of industrial design (APCI)[1] tabled the notion of durable development as one of its objectives. Every year, the "Observeur du design" event, which it has been organizing since 1999, presents awards to a selection of the finest achievements in the area that take into account "eco-design"—that is, the attention afforded the environment in product design in endeavoring to limit negative impacts at every stage of the article's life (from the extraction of the raw materials, through manufacture, distribution, and utilization to the value accrued from its waste). The "Observeur du design" event, held in partnership with the French agency for the environment and energy control (ADEME), awards a special prize intended to increase sensitivity to environmental protection among design professionals. In 2004, the fifth such event selected 165 products, of which thirty-one were awarded a "star," while the ADEME allotted its blue ribbon to several endeavors, notably: to a fuel cell, the Roller Pac, running on hydrogen; quiet, compact, and emitting only water vapor, a forerunner of the power-generating units of tomorrow; to an office chair, Please 2, whose various improvements are accompanied by an "eco-profile," a

1 Initially established in 1983 on the initiative of the French Ministries of Industry and of Culture, ten years afterwards the APCI became a private institution.

quantitative specification of its environmental data; and to a transparent pocket, the Bio-File, an aid in the selective sorting of office waste. Manufactured from a biomaterial made from cornstarch, this last replaces oil-based plastics and can be recycled directly or turned to compost in the same way as the paper it contains. "Eco-design" underlines the point that not only are the initial stages of a product crucial environmentally, but that concern for the environment can become a source of creativity and innovation.

Whereas at one time priority was given to the treatment of distribution, brains are presently challenged by inventory control: certain companies are now striving to recover materials, becoming concerned with avoiding irreversible transformations and embracing a desire to rehabilitate contaminated environments both as a sign of ecological awareness and responsibility and as a way of locking into new attitudes to profitability. For growing environmental focus is a factor far from indifferent to consumer demand, as well as contributing to the preparation of tomorrow's markets. It is in the area of packaging that producers, researchers, managers, and designers work most closely together in developing solutions based sometimes on biodegradability, sometimes on materials that, though still containing harmful substances, can be realistically collected and recycled. Hence the seminal importance of operational research focusing on continuously monitoring a product through its entire lifecycle.

The Institut National de Design Packaging (INDP) in Cognac, in western France, for instance, now offers to the various professionals concerned contacts, diagnoses, and resources. The aim of the INDP is to encourage creativity and development throughout the "Atlantic arc," plugging into two complementary factors in the regional economy—the well-established position of the printing, graphic art, and communication sectors, and a packaging industry that represents, for example, a quarter of all business activity in the Poitou-Charentes region. Its director, Jean-Christophe Boulard, highlights the extent to which the packaging industry is embracing the integration of ecological demands at every stage of product lifecycle more and more comprehensively. For example, the Otor cardboard factory in Cognac employs 90 percent recycled paper for the manufacture of corrugated cardboard, and the food conditioning manufacturer Ameor at Barbezieux has replaced rigid paperboard packing with printed flexible packing that is used to both contain and present the product.

ROLLER PACK TRANSPORTABLE COMBUSTIBLE BATTERY BY THE FIRM OF AXANE (AIR LIQUIDE), AWARDED A SPECIAL CITATION BY THE ADEME (FRENCH AGENCY FOR THE ENVIRONMENT AND ENERGY CONTROL), *OBSERVEUR DU DESIGN*, FRANCE, 2004.

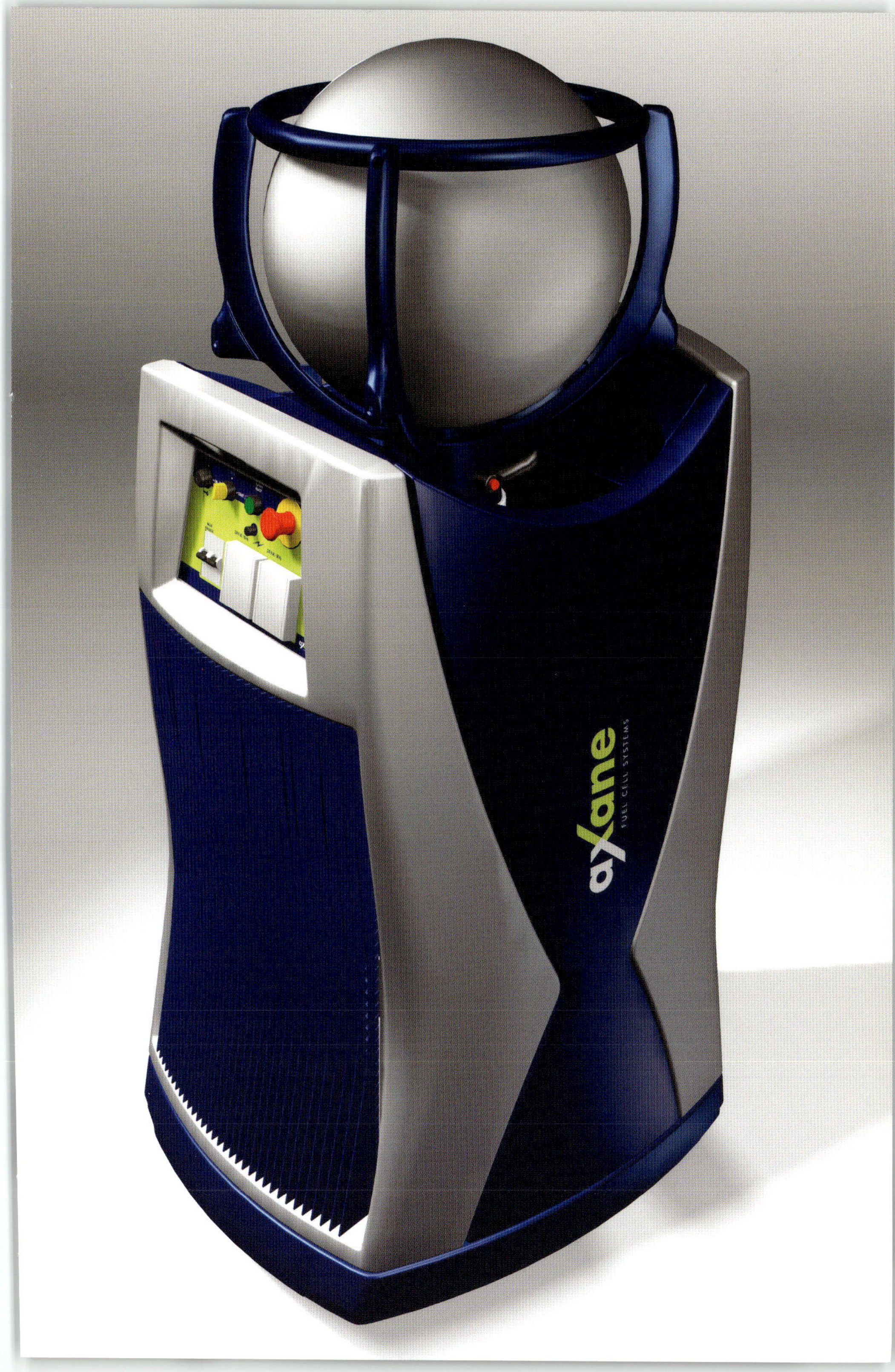
axane
FUEL CELL SYSTEMS

The general public's greater attention to the packaging sector is explained by the widespread introduction of waste sorting: "The practice of solid waste selection," Jean-Christophe Boulard explains, "has raised awareness and changed expectations among consumers. In France today, 42 million people sort their garbage and thus have shifted from disposable to recyclable, and they expect from industry uncomplicated, mono-material packaging corresponding to clearly identifiable recycling channels."[2]

Some designers recommend returning to natural materials, one such being Andrea Branzi with his series of Domestic Animals made of branches of trees, while others favor using only one, readily recyclable material exclusively, such as Philippe Starck in his Bubble settee. Needs must: during the Occupation in France, haute couture turned wooden soles and Vichy fabrics to its advantage. Ecological design is compelled to make a virtue of need—that is, it has to create style and art from natural materials, or, if not, at the very least from biodegradable or entirely recyclable ones.

In their sales points, car manufacturers now stress that more than 80 percent of the components of their product can be recycled, an argument that carries considerable weight in countries where Green parties and ecologist movements exert a significant influence on public opinion. By the end of 2000, there had had been lodged in Europe 9,663 and in Japan 3,990 certificates of the ISO 14001 standard, which warrants a system of "durable management" coupled to a commitment to constant improvement.

Even in the United States, which has a well-earned reputation for being the greatest waster of energy in the world, the pressure of public opinion has led to anti-pollution legislation governing recycling and waste reduction. In a book that wears its optimism on its sleeve, Michel Drancourt underlined the extent to which "managerial methods required for achieving quality echo those needed for sustainable development. One finds the same striving for economy (raw materials, energy, transport) and the creative imagination to improve products and services and to innovate."[3] He also refers to the many American firms concerned with scoring highly on the "social rating" carried out by several non-profit-making associations such as

PLEASE 2 CHAIR, WINNER OF A SPECIAL AWARD FROM THE ADEME, *OBSERVEUR DU DESIGN*, FRANCE, 2004. DESIGNER: GLEN OLIVER LÖW FOR STEELCASE, SA.

2 See "Un design aux exigences environnementales," *L'actualité Poitou-Charente*, no. 68, April-May-June, 2005, pp. 20-21.
3 M. Drancourt, *Les arbres ne poussent pas jusqu'au ciel...donc il faudra beaucoup de forêts – Faut-il réinventer le progrès?* , Paris: Pearson Education, 2003.

the Center for Responsibility in Business, the Domini 400 Social Index, or the Dow Jones Sustainability Index, which factor in indicators relating to environmental policy. Even large and inevitably polluting multinational corporations (Shell, BP, Lafarge) intend to build in developmental objectives directed toward reducing pollution, improving production technologies, reducing energy use, and soil rehabilitation. The growing clout wielded by standards and certificates covering "environmental values" is undeniable and surely reflects the commitment of a number of firms to equip themselves with the means of contributing to the "Green" struggle, and some have even been prepared to join non-governmental organizations (NGOs) such as the World Resource Institute in undertaking studies and evaluations measuring the progress made in this new approach to economic value.

In the context of his resolutely liberal concept of economics, Michel Drancourt encapsulates the three principles that have to underpin any such approach: first, more has to be manufactured with less; second, which he calls "transversal," the search for quality has to systematic; third, that the concept of duration has to be reintroduced into consumption and into the economy generally. It goes without saying that design, in compatibility with such principles, is to intervene at every level of development—manufacture, product management, processes, machines, and networks of all kinds. This does not mean, however, that "ecological design" and the "Green" goodwill shown by certain companies will be enough to meet the environmental challenges faced by the planet.

A report published by the Organization for Economic Co-operation and Development (OECD) on "voluntary approaches in environmental policy" shows just how limited their overall effectiveness has proved: on the one side, there is opportunistic behavior by certain companies, which the report characterizes as environmental policy "stowaways," in that they manage to avoid tax or stricter regulations without seeking to take on the wider constraints; on the other, a range of competitive pressures, like the threat of unemployment or relocation, are such that a given firm's commitment to the environment is bound to appear "half-hearted" in the sense that in the short run it has to do its utmost to avoid compromising the bottom line.[4] It is revealing that

4 *Les approches volontaires dans les politiques de l'environnement: Efficacité et combinaison avec d'autres instruments d'intervention*, OCDE, 2003.

the OECD, a champion if ever there was one of the free-market economy and of liberal practices, has so few illusions in this connection, concluding that "economic instruments on an economic scale"—i.e., a more resolute and direct intervention on behalf of governments—"may constitute a better policy choice than voluntary approaches from the point of view of environmental effectiveness as well as of governmental efficiency."[5] As in the case of the fight against poverty and galloping social inequality, such statements amount to highlighting the limits of private action and non-profit institutions compared to the need for the intervention of public bodies in advancing a social policy: it is as if the OECD is on the brink of recommending that the welfare state extend its remit over ecological welfare. By the same token though, it shows how much an economic or political (or ideological) stance can condition responses to the present situation. For some, only the market and spontaneous or voluntary intervention of economic entities are capable of responding to the conditions under which companies may find it beneficial to implement an ecological strategy. For others, a policy of laissez-faire, like that of undirected innovation, can result only in raising the bar of what is considered intolerable pollution, and therefore what is needed is direct state intervention, from regulatory constraints to controls and sanctions.

One can certainly imagine policies implementing more stringent, i.e., more restrictive, checks and balances that would promote not so much the adaptation of existing technologies than the development of new ones firmly conditioned by environmental objectives and directed toward them. Without wanting to venture too optimistically into futurology, it seems that, in three fields at least, innovation in materials and functions today presents designers with the chance to participate in a more persuasive reconfiguration of the entire economy: first of all there is the virtual realm, where digital techniques offer the possibility of creating and recreating ad infinitum worlds free of "environmental fallout"; then come new composite materials (epoxy resins, polyesters in conjunction with fiberglass, carbon, boron, aramid, etc.), which pave the way to compounds with properties inconceivable in traditional materials; finally, the less developed yet exceptionally promising field of nanotechnology. The salient characteristic of composites was to integrate various functions into their composition, in the process increasing their potential

5 Ibid., p. 17.

for interfacing with the end user: this was the onset of a mode of CAD and manufacture of a modular material that makes it possible to identify technical needs a priori and to develop *ex post* a material specifically adapted to them. In nanotechnology, the convergence of the information sciences, biotechnology, and the cognitive sciences promises to go much further in substituting technical artifacts with a production process in direct competition with nature.

If one believes Eric Drexler, who in his Foresight Institute at Palo Alto has taken on the mantle of a prophet, the future lies in artificial nanomechanics that mimic biological self-organizing principles, where one will be able ask anything and everything of nano-networks, nano-assemblers, and nano-machines driven by nano-molecular energy[6]. On the basis of material with microscopic components, whose conception and functions will rely primarily on the algorithms of molecular electronics, the designer's task would amount to furnishing the final algorithm with the aesthetic best adapted to conquering the market. The question is whether actually to reproduce all the processes of self-organization in the living world by doing as nature does, or, if possible, better, as projected in a plan richly endowed with federal funds that appears in a report by the National Science Foundation beneath a title that says it all: *Converging Technologies for Improving Human Performance*.[7] The promise is nothing less than the unification of the sciences and technology, universal material and spiritual wellbeing, world peace, an access to inexhaustible sources of energy, and, of course, an end to all concern with environmental pollution. The convergence of scientific disciplines, the manipulation of inert and living materials on the molecular scale, construction and reconstruction of new functional structures atom by atom—it is easy to imagine the role designers could fulfill at every stage of this new type of manufacturing process in devising products tailor-made to appeal to the market. After the era of art playing its part in industry, comes the era of technology reproducing the very art of nature. But we are not yet at the stage of the industrial triumph of the nano-scale. On the contrary, it seems that even now there is ample scope

6 See E. Drexler, *Engines of Creation: The Coming Era of Nanotechnology*, New York: Anchor Books, 1986. (available online on the site of *Foresight Institute*, www.foresight.org). A nanometer is a millionth of a millimeter. **7** NSF, Washington, 2002. The report is accessible on the Internet, http://wtec.org/ConvergingTechnologies.

OIL SLICK IN BRITTANY FROM
THE STRICKEN LIBERIAN TANKER
AMOCO CADIZ, MARCH 1972.

NOKIA

for worrying, as Jean-Pierre Dupuy has shown, about the "unparalleled risks" Eric Drexler's program involves, if by any chance everything he envisages as possible in the medium term was indeed put into practice.[8]

There is, for instance, the threat of "global ecophagia," the uncontrolled auto-replication of nanotechnology following a programming accident in which "whole or a part of the biosphere would be destroyed through self-reproduction by nano-machines exhausting carbon reserves." Jean-Pierre Dupuy seems to take such risks seriously when he writes that he "can only really frighten those who believe in the possibility of such machines. It's enough to deny this possibility to throw out the pseudo-risk with a shrug of the shoulders." Patience is called for then. Before designers can be closely and directly associated with the nanotechnology market, the promise held by the "fabulous innovations" resulting from the convergence of all scientific disciplines, whose triumphs researchers are already fantasizing about, have to be fulfilled—and without these same innovations confirming the fears that they have already instilled in the eyes of people like Jean-Paul Dupuy: after the Cartesian dream of "being master over and proprietor of nature," there comes the impossibility of "mastering the mastery" of the artificial worlds human genius has spawned.

We have here reached a point well beyond design and the exuberant optimism to which Raymond Loewy testified in his visionary tribute on the eve of the boom period. Faced with the uncertainties and threats that vex the beginning of the twenty-first century, the political climate can hardly be propitious to ecology.

The decisions called for are quite simply not taken on the scale required since, justifiably or not, other priorities always cloud the issue. On the contrary, the 21st century opened against a backdrop of crisis and risk. So much so that disaster has itself become an art object, as was illustrated in Paris at the exhibition "Ce qui arrive" ("What is happening") organized in 2003 at the Fondation Cartier by Paul Virilio.[9]

MOBILES: THE ENGLISH COMPANY SMILE PLASTICS SPECIALIZES IN RECYCLING PLASTICS AND MARKETS DECORATIVE PANELS MANUFACTURED AT THE FOUNDRY FROM OUT-OF-COMMISSION CONSUMER GOODS (HERE, TELEPHONES).

8 J.-.P. Dupuy, "Quand les technologies convergeront," *La Jaune et la Rouge* (the review of the Polytechniques, Paris), no. 500, pp. 47-52, reprinted in *Futuribles*, no. 300, September 2004, pp. 5-18. **9** See my review "Au musée des désastres," *Futuribles*, no. 284, March 2003, 65-76.

Two years later, on the initiative of MoMA (the Museum of Modern Art), New York, three hundred design objects were assembled in an effort to describe the countless ways in which post-modernity tries to face up to its everyday angst. The curators had initially thought of centering what is in fact the first design show since the museum reopened on the theme of fear, settling finally for the marginally less negative and discouraging one of risk and security, under the umbrella title "SAFE: Design Takes on Risk." The trauma of September 11 clearly emerges in the six sections it explored: shelter, armor, property, the every-day, emergency, and awareness, as if the sole vocation of a young generation of architects and designers taking part is to protect humanity from the threats that beset it, ranging from terrorism to natural disaster.

Like lessons from the history of current affairs and in a brew that seemed to verge on the derisive, the show featured ingenious shelters for the poor, refugees and victims of wars or genocide, nests to ensure the emotional well-being of babies, luminous blankets for overcoming the fear of the dark, barbed wire with points in the shape of butterflies, and a figure in the shape of a mushroom cloud "for fragile people in these angst-ridden times."[10].

As society struggles to come to terms with risk, it is clear that designers have the potential, or perhaps even the mission, to contribute without delay to the advent of a new historical stage in the Industrial Revolution, which must necessarily strive (if catastrophe is to be avoided) to reconcile the machine and its products with environmental imperatives.

After that at the Fondation Cartier, the MoMA show is just one more curtain-raiser to the reshaping of a world "outside history" that can only dream of freeing itself from fear and of defusing every danger.

ANTHONY DUNNE, FIONA RABY, MICHAEL ANASTASSIADES, ATOMIC MUSHROOM, A MUSHROOM FOR FRAGILE PEOPLE IN ANGST-RIDDEN TIMES, AT THE EXHIBITION *SAFE: DESIGN TAKES ON RISK*, MUSEUM OF MODERN ART, NEW YORK, 2005.

10 Cf. "Quand nos peurs inspirent les créateurs: Saint Design protégez-nous!" *Le Monde* 2, 24/12/05, 52-56.

BIOGRAPHIES

RAYMOND GUIDOT IS ONE OF THE CHIEF SPOKESMEN ON DESIGN IN FRANCE. AN ENGINEER AT THE ÉCOLE NATIONALE SUPÉRIEURE DES ARTS ET MÉTIERS, FROM 1961 TO 1969 HE COLLABORATED WITH DESIGNER ROGER TALLON WITHIN THE AMBIT OF THE TECHNÈS AGENCY. SINCE 1972, HE HAS BEEN RESPONSIBLE FOR ORGANIZING EXHIBITIONS AT THE CENTRE DE CRÉATION INDUSTRIELLE FOUNDED IN 1969 AT THE MUSÉE DES ARTS DÉCORATIFS IN PARIS, BEFORE IT BECAME A DEPARTMENT OF THE POMPIDOU CENTER, PARTICIPATING ON ITS EXHIBITION COMMITTEES. AS A TEACHER, HE HAS BEEN ACTIVE AT THE ÉCOLE NATIONALE SUPÉRIEURE DES ARTS DÉCORATIFS, PARIS, THE ÉCOLE NATIONALE SUPÉRIEURE DE CRÉATION INDUSTRIELLE, PARIS, AND AT STRATE COLLEGE, ISSY-LES-MOULINEAUX. TODAY, HE WORKS AT THE INSTITUT SUPÉRIEUR DES ARTS AND AT THE ÉCOLE CANTONALE DES ARTS IN LAUSANNE. AUTHOR OF MANY ARTICLES, HE HAS ALSO PUBLISHED *DES ARTS PLASTIQUES AUX ARTS APPLIQUÉS* (NO. 85 OF THE CENTRE NATIONAL DE DOCUMENTATION PÉDAGOGIQUE, 1992) AND *HISTOIRE DU DESIGN, 1940–2000* (PARIS: HAZAN, 2000).

JEAN GRENIER IS AN INDUSTRY ENGINEER AND GRADUATE OF THE CENTRE NATIONAL DES ARTS ET MÉTIERS (CNAM), PARIS. HE IS A SPECIALIST IN MATERIALS, PARTICULARLY IN SYNTHETIC SUBSTANCES AND THEIR TRANSFORMATIONS. HE HAS SPENT MANY YEARS STUDYING, DESIGNING, AND OVERSEEING THE INDUSTRIALIZATION OF MORE THAN 800 PRODUCTS AND APPROXIMATELY 15,000 TECHNICAL COMPONENTS IN THE MOST VARIED FIELDS (ELECTRIC HOUSEHOLD APPLIANCES, TOYS, COMPUTING, LUXURY GOODS, TRANSPORT, ETC.). HIS INDUSTRIAL CULTURE IS UNIVERSAL INSOFAR AS THE STUDIES HE HAS UNDERTAKEN FOR MANY COMPANIES HAVE RESULTED IN ARTICLES PRODUCED THE WORLD OVER. AT THE SAME TIME, HE HAS BECOME AN AUTHORITATIVE HISTORIAN OF INDUSTRIAL PROCESSES. IN 1979, HE FOUNDED THE ENGINEERING AND DESIGN BUREAU, SOLUS CONSEIL, WHICH, IN 1994, BECAME GRENIER CONCEPT. HE HAS BEEN TEACHING TECHNOLOGY AT THE ÉCOLE NATIONALE SUPÉRIEURE DES ARTS DÉCORATIFS, IN PARIS SINCE 1972 AND, FOR THE LAST TEN YEARS, AT CRÉAPOLE, PARIS. HIS PASSION FOR COMMUNICATING HIS KNOWLEDGE HAS LED HIM TO COLLABORATE ON SEVERAL BOOKS ON DESIGN AND TO SUPERVISE A LARGE NUMBER OF GRADUATION PAPERS AND THESES, IN PARTICULAR THE TEN VOLUME *ENCYCLOPÉDIE DES MATÉRIAUX*, A DOCTORATE THESIS PRESENTED IN ATHENS IN 1989 BY HÉLÈNE MITROPOLOU.

"INTERNATIONAL CONGRESS ON COMPOSITE MATERIALS TODAY AND TOMORROW" (BOTTOM RIGHT, FROM LEFT TO RIGHT, JEAN ATTIC AND RAYMOND GUIDOT), LYON, SEPTEMBER 1971.

AFTER TRAINING AS A MATHEMATICIAN, FROM THE 1970S **JEAN-BAPTISTE TOUCHARD** PARTICIPATED IN THE VERY BEGINNINGS OF MULTIMEDIA NETWORK COMPUTING. IN 1978, WHILE HEAD OF THE ATELIER DE RECHERCHE ET DE CRÉATION SCIENTIFIQUE DE L'INSTITUT NATIONAL DE L'AUDIOVISUEL, HE NOTABLY CREATED THE FIRST FOUR NETWORK GAMES FOR MINITEL AND PRODUCED THE FIRST ANIMATED SYNTHESIZED IMAGES TO BE BROADCAST BY TELEVISION. AFTER SETTING UP THE FIRST EUROPEAN MASTER'S DEVOTED TO MULTIMEDIA AS DIRECTOR OF RESEARCH AT THE ÉCOLE NATIONALE SUPÉRIEURE DE CRÉATION INDUSTRIELLE, IN 2000 HE FOUNDED TXTNET.COM. IN 2003, HE JOINED STRATE COLLEGE IN ISSY-LES-MOULINEAUX, FRANCE, AND IS TODAY DIRECTOR OF ITS MASTER'S IN DI&CI ("INDUSTRIAL DESIGN AND INNOVATION MANAGEMENT"). HE HAS WRITTEN FOURTEEN WORKS, MANY ARTICLES, AND SEVERAL SOFTWARE PRODUCTS. HE HAS ALSO BROUGHT OUT FOUR AUDIOVISUAL PACKAGES AND PRODUCED A DOZEN INTERACTIVE MULTIMEDIA TITLES AS WELL AS A SIMILAR NUMBER OF INTERNET SITES.

JEAN-JACQUES SALOMON IS PROFESSOR EMERITUS AND, SINCE 1978, HAS OCCUPIED THE CHAIR OF TECHNOLOGY AND SOCIETY AT THE CONSERVATOIRE NATIONAL DES ARTS ET MÉTIERS (CNAM), PARIS, IN WHICH HE SET UP THE SCIENCE, TECHNOLOGY, AND SOCIETY CENTER. HOLDING A DOCTORATE IN PHILOSOPHY AND HISTORY OF SCIENCE, HE FOUNDED AND, FROM 1963 TO 1983, DIRECTED THE POLITICS OF SCIENCE AND TECHNOLOGY DIVISION AT THE ORGANIZATION FOR ECONOMIC COOPERATION AND DEVELOPMENT (OECD). HE WAS THE COFOUNDER WITH DEREK DE SOLLA PRICE OF THE INTERNATIONAL COUNCIL FOR SCIENCE POLICY STUDIES AND PRESIDENT OF THE COLLÈGE DE LA PRÉVENTION DES RISQUES TECHNOLOGIQUES (CPRT) IN PARIS BETWEEN 1991 AND 1996. HE IS A MEMBER OF THE ACADEMY OF SCIENCE, NEW YORK, AND HAS TAUGHT AT MIT, HARVARD, MONTREAL, SÃO PAULO, AND IN INDIA.

SELECTIVE BIBLIOGRAPHY

ABBATE, JANET.
INVENTING THE INTERNET.
CAMBRIDGE, MASS.: MIT PRESS, 1999.

ALDERSEY-WILLIAMS, HUGH.
NEW AMERICAN DESIGN: PRODUCTS AND GRAPHICS FOR A POST-INDUSTRIAL AGE.
NEW YORK: RIZZOLI, 1988.

LES ANNÉES PLASTIQUES (exh. cat.).
PARIS: ÉDITIONS ALTERNATIVES/LA VILLETTE, 1986.

BAKER. FIONA, BAKER, KEITH,
20TH CENTURY FURNITURE.
LONDON: CARLTON, 2000.

BEAUREPAIRE, EMANUEL.
RECYCLAGE DES PRODUITS EN FIN DE VIE, GESTION: TECHNIQUES-RESPONSABILITÉS.
PARIS: AFNOR, 2003.

BARRÉ-DESPOND, ARLETTE (gen. ed.).
DICTIONNAIRE INTERNATIONAL DES ARTS APPLIQUÉS ET DU DESIGN.
PARIS: ÉDITIONS DU REGARD, 1996.

BAUDET, JEAN.
DE LA MACHINE AU SYSTÈME, HISTOIRE DES TECHNIQUES DEPUIS 1800.
PARIS: VUIBERT, 2004.

BENT WOOD AND METAL FURNITURE, 1850-1946.
NEW YORK: THE AMERICAN FEDERATION OF ARTS, 1987.

BOISSIÈRE, OLIVIER.
STREAMLINE: LE DESIGN AMÉRICAIN DES ANNÉES 30.
PARIS: RIVAGES, 1987.

BONY, ANNE.
LES ANNÉES 50.
PARIS: ÉDITIONS DU REGARD, 1982.

FURNITURE AND INTERIORS OF THE 1960S.
PARIS: FLAMMARION, 2004.

FURNITURE AND INTERIORS OF THE 1970S.
PARIS: FLAMMARION, 2005.

LES ANNÉES 30.
PARIS: ÉDITIONS DU REGARD, 1987.

LES ANNÉES 10.
PARIS: ÉDITIONS DU REGARD, 1991.

LES ANNÉES 70.
PARIS: ÉDITIONS DU REGARD, 1993.

LES ANNÉES 80.
PARIS: ÉDITIONS DU REGARD, 1995.

BOURGEOIS, RENÉ, with JACQUES KESSLER and HENRI CHAUVEL.
MEMOTECH, GÉNIE DES MATÉRIAUX.
PARIS: CASTEILLA (EL ÉDUCATIVE), 2001.

BRANZI, ANDREA.
THE HOT HOUSE: ITALIAN NEW WAVE DESIGN.
CAMBRIDGE, MASS.:MIT PRESS, 1983.

FOURTH METROPOLIS: DESIGN AND ENVIRONMENTAL CULTURE.
NEW YORK:ART PUBLISHING INC., 1991.

NOUVELLES DE LA MÉTROPOLE FROIDE: DESIGN ET SECONDE MODERNITÉ.
PARIS: ÉDITIONS DU CENTRE POMPIDOU (COLL. "LES ESSAIS"), 1988.

BRANZI, ANDREA.
IL DESIGN ITALIANO, 1964-1990.
MILAN: ELECTA, 1996.

BRUNHAMMER, YVONNE.
LES STYLES DES ANNÉES 30 À 50.
PARIS: ÉDITIONS DE L'ILLUSTRATION BASCHET, 1987.

ART DECO STYLE.
LONDON: ST. MARTIN'S PRESS, 1984

BYARS, MEL.
THE DESIGN ENCYCLOPAEDIA.
LONDON: LAURENCE KING, 1994.

50 LIGHTS: INNOVATION IN DESIGN AND MATERIALS.
CRANS-PRÈS-CELIGNY: ROTOVISION, 1997.

50 TABLES: INNOVATION IN DESIGN AND MATERIALS.
CRANS-PRÈS-CELIGNY: ROTOVISION, 1997.

50 CHAIRS: INNOVATION IN DESIGN AND MATERIALS.
CRANS-PRÈS-CELIGNY: ROTOVISION, 1997.

ON OFF NEW ELECTRONIC PRODUCTS.
LONDON: LAWRENCE KING PUBLISHING, 2003.

BYARS, MEL, BARRÉ-DESPOND, ARLETTE.
CENT OBJETS: UN SIÈCLE DE DESIGN. PARIS: LES ÉDITIONS DE L'AMATEUR, 1999.

CASCIANI, STEFANO.
FURNITURE AS ARCHITECTURE: DESIGN AND ZANOTTA PRODUCTS.
MILAN: ARCADIA EDIZIONI, 1988.

CASCIANI, STEFANO.
100 CHEFS-D'ŒUVRE DE LA COLLECTION DU VITRA DESIGN MUSEUM.
WEIL AM RHEIN: VITRA DESIGN MUSEUM, 1996.

COLIN, CHRISTINE.
DESIGN AUJOURD'HUI.
PARIS: FLAMMARION, 1988.

DARMON, FRANÇOISE.
DU SENS DANS L'UTILE.
PARIS, ÉDITIONS DU CHÊNE, 1992.

DESIGN 1950-2000.
PARIS: FLAMMARION/MUSÉE DES ARTS DÉCORATIFS, MONTREAL, 2000.

DESIGN JAPONAIS, 1950-1995 (exh. cat.).
PARIS: ÉDITIONS DU CENTRE POMPIDOU, 1996.

DICTIONNAIRE ILLUSTRÉ DES DESIGNERS DU XXE SIÈCLE.
PARIS: CELIV, 1992.

DARMON, FRANÇOISE, with THOMAS DE GALIANO and MICHEL RIVAL.
DICTIONNAIRE DES INVENTEURS ET INVENTIONS. PARIS: LAROUSSE/BORDAS, 1996.

DORMER, PETER.
DESIGN SINCE 1945.
LONDON: THAMES AND HUDSON, 1993.

DRANCOURT, MICHEL.
LES ARBRES NE POUSSENT PAS JUSQU'AU CIEL... DONC IL FAUT BEAUCOUP DE FORÊTS. FAUT-IL RÉINVENTER LE PROGRÈS?
PARIS: PEARSON EDUCATION, 2003.

DREYFUSS, HENRY.
DESIGNING FOR PEOPLE.
HARMONDSWORTH: PENGUIN, 1967 (1st ed. 1955).

DROSTE, MAGDALENA, BAUHAUS ARCHIV,
BAUHAUS 1919–1933.
COLOGNE: TASCHEN, 1993.

FAYOLE, CLAIRE.
LE DESIGN. PARIS: ÉDITIONS SCALA (COLL. "TABLEAUX CHOISIS"), 1998.

C'EST QUOI LE DESIGN?,
PARIS: AUTREMENT JUNIOR ARTS, 2002.

FEHRMAN, CHERIE, FEHRMAN, KENNETH.
POSTWAR INTERIOR DESIGN, 1945-1960. NEW YORK: VAN NOSTRAND REINHOLD, 1987.

FIEDLER, JEANNINE, FEIERABEND, PETER.
BAUHAUS. COLOGNE: KÖNEMANN, 1999.

FIELL, CHARLOTTE, FIELL, PETER.
MODERN FURNITURE CLASSICS.
LONDON: THAMES AND HUDSON, 2001.

MODERN CHAIRS.
COLOGNE: TASCHEN, 1993.

1000 CHAIRS.
COLOGNE: TASCHEN, 1997.

DESIGNING THE 21ST CENTURY.
COLOGNE: TASCHEN, 2001.

GANTZ, CARROLL.
100 YEARS OF DESIGN: A CHRONOLOGY, 1895-1995.
NEW YORK: INDUSTRIAL DESIGNERS SOCIETY OF AMERICA, 1999.

GAY, DANIEL.
MATÉRIAUX COMPOSITES.
PARIS: HERMÈS, 1997.

GUIDOT, RAYMOND.
HISTOIRE DU DESIGN, DE 1940 À NOS JOURS.
PARIS: HAZAN, 1999.

GUIDOT, RAYMOND (ed.).
DESIGN CARREFOUR DES ARTS.
PARIS: FLAMMARION, 2003.

GUILLAUME, VALÉRIE, with BENOÎT HEILBRUNN and OLIVIER PEYRICOT.
ABCDAIRE DU DESIGN.
PARIS: FLAMMARION, 2003.

ÉRIARD-DUBREUIL, BERTRAND,
IMAGINAIRE TECHNIQUE ET ÉTHIQUE SOCIALE, ESSAI SUR LE MÉTIER D'INGÉNIEUR.
BRUSSELS: DE BOECK ET LARCIER, 1997.

HESLETT, JOHN.
INDUSTRIAL DESIGN.
LONDON: THAMES AND HUDSON, 1987.

ICONS OF DESIGN: THE 20TH CENTURY. MUNICH/LONDON/NEW YORK: PRESTEL, 2000.

ITALY THE NEW DOMESTIC LANDSCAPE (exh. cat.). NEW YORK: THE MUSEUM OF MODERN ART, 1972.

JACOMY, BRUNO.
UNE HISTOIRE DES TECHNIQUES.
PARIS: ÉDITIONS DU SEUIL, 1990.

JAPANESE DESIGN.
COLOGNE: TASCHEN, 1994.

JOLLANT-KNEEBONE, FRANÇOISE.
LA CRITIQUE EN DESIGN: CONTRIBUTION À UNE ANTHOLOGIE.
PARIS: ÉDITIONS JACQUELINE CHAMBON, 2003.

JOUNOT, ALAIN.
LE DÉVELOPPEMENT DURABLE.
PARIS: AFNOR, 2004.

JOUSSET, MARIE-LAURE, POTSCH, KATARINA V.
ALEXANDER VON VEGESACK: PORTRAIT D'UNE COLLECTION.
PARIS: ÉDITIONS DU CENTRE POMPIDOU (COLL. "JALONS"), 1993.

KATZ, SYLVIA.
LES PLASTIQUES, DE LA BAKÉLITE AU HIGH-TECH.
PARIS: DENOËL, 1985.

KAZAZIAN, THIERRY, HENRY, MATHILDE, LESQUOY, ANNE-LAURE, BENBOUALI, SANDRINA, SAPORTA, HENRI.
LE CYCLE DE L'EMBALLAGE.
PARIS: MASSON, 1995.

KJELLBARG, PIERRE,
LE MOBILIER DU XXe SIÈCLE: DICTIONNAIRE DES CRÉATEURS.
PARIS: LES ÉDITIONS DE L'AMATEUR, 1994.

LAURENT, STÉPHANE.
CHRONOLOGIE DU DESIGN.
PARIS: FLAMMARION, 1999.

LEROI-GOURHAN, ANDRÉ.
L'HOMME ET LA MATIÈRE.
PARIS: ALBIN-MICHEL (COLL. "SCIENCES D'AUJOURD'HUI"), 1943, 1971.

MILIEUX ET TECHNIQUES.
PARIS: ALBIN-MICHEL (COLL. "SCIENCES D'AUJOURD'HUI"), 1945, 1973.

GESTURE AND SPEECH.
CAMBRIDGE, MASS.: MIT PRESS, 1993.

LOEWY, RAYMOND,
NEVER LEAVE WELL ENOUGH ALONE.
BALTIMORE: JOHNS HOPKINS PRESS, 2002.

MATERIO.
MATERIAL WORLDS 2, INNOVATIVE MATERIALS FOR ARCHITECTURE AND DESIGN.
BASEL: BIRKHAÜSER, 2006.

MCDERMOTT, CATHERINE.
20TH CENTURY DESIGN: DESIGN MUSEUM. LONDON: CARLTON, 1997.

MANZINI, EZIO.
ARTEFACTS: VERS UNE ÉCOLOGIE DE L'ENVIRONNEMENT ARTIFICIEL,
PARIS: ÉDITIONS DU CENTRE POMPIDOU (COLL. "LES ESSAIS"), 1991.

THE MATERIAL OF INVENTION, MATERIALS AND DESIGN.
NEW YORK: HYPERION, 1989.

MEMPHIS: THE NEW INTERNATIONAL STYLE.
MILAN: ELECTRA, 1981.

NAYLOR, GILLIAN
THE BAUHAUS REASSESSED: SOURCES AND THEORY.
NEW YORK: E.P. HUTTON, 1985.

De NOBLET, JOCELYN.
DESIGN: LE GESTE ET LE COMPAS
PARIS: SOMOGY, 1988.

De NOBLET, JOCELYN (gen. ed.).
DESIGN: MIROIR DU SIÈCLE (exh. cat.).
PARIS: GRAND PALAIS/ FLAMMARION, 1993.

L'OBJET CRÉÉ PAR L'HOMME.
BRUSSELS: ÉDITIONS DE LA CONNAISSANCE, 1968.

PAGE, MARIAN.
FURNITURE DESIGNED BY ARCHITECTS.
LONDON:THE ARCHITECTURAL PRESS, 1980.

PAPANEK, VICTOR,
DESIGN FOR A REAL WORLD,
CHICAGO: ACADEMY CHICAGO PUBLISHERS, 1985

PAQUET, G.
GUIDE DE L'USINAGE.
PARIS: DELAGRAVE, 2000.

PULOS, ARTHUR J.
AMERICAN DESIGN ETHIC: A HISTORY OF INDUSTRIAL DESIGN TO 1940.
CAMBRIDGE, MASS.: MIT PRESS, 1983.

QUARANTE, DANIELLE.
ÉLÉMENTS DE DESIGN INDUSTRIEL.
PARIS: POLYTECHNICA, 1994 (2nd ed.).

REYNE, MAURICE.
LE COMPOSITE.
PARIS: PUF, 1995.

LES MATÉRIAUX NOUVEAUX.
PARIS: HERMÈS, 1990.

SALOMON, JEAN-JACQUES.
LE SCIENTIFIQUE ET LE GUERRIER.
PARIS: ÉDITIONS DU SEUIL, 2001.

SPARKE, PENNY.
A CENTURY OF DESIGN:

DESIGN PIONEERS OF THE 20TH CENTURY.
LONDON: MITCHELL BEAZLEY, 1998.

TANIZAKI, JUNICHIRO,
IN PRAISE OF SHADOWS.
LEETES ISLANDS BOOKS, 1980.

TOUCHARD, JEAN-BAPTISTE.
IMAGES NUMÉRIQUES.
PARIS: CEDIC/NATHAN, 1987.

L'UTOPIE DU TOUT PLASTIQUE, 1960-1973 (exh. cat.).
BRUSSELS: FONDATION POUR L'ARCHITECTURE, 1994.

WHITFORD, FRANK.
*BAUHAUS.*LONDON: THAMES AND HUDSON, 1984.

REVIEWS AND PERIODICALS,

ARCHICRÉÉ, ARTPRESS, AZIMUTS, DESIGN ENTRÉE EN MATIÈRES, INTRAMUROS, L'USINE NOUVELLE, SCIENCE ET VIE, FRANCE, *BLUEPRINT,* GREAT BRITAIN, *ABITARE, DDN, DOMUS, MODO,* ITALY.

ACKNOWLEDGMENTS AND PHOTOGRAPHIC CREDITS

I WOULD LIKE TO THANK ALL THOSE WHO HAVE MADE THE PUBLICATION OF THIS WORK POSSIBLE, PARTICULARLY JULIE ROUART AND CLÉMENT DIRIÉ AT ÉDITIONS FLAMMARION, WHOSE INTEREST IN THE SUBJECT MATTER AND PROFESSIONALISM WERE INSTRUMENTAL TO THE SUCCESS OF THE JOB IN HAND.

ALSO DIANE SAUNIER, FRANÇOIS CHÉRY, FRANÇOIS GUIDOT, AND ALAIN POUPART WHO, BY THEIR ADVICE AND CARE TAKEN IN THE READING OF THE TEXT, CONTRIBUTED TO THE FINAL PRODUCT.

TO LINE MARTIN THE DESIGN OF THE BOOK, AND ALSO THOSE WHO, WHETHER IN A PROFESSIONAL CAPACITY—FROM DASSAULT, ALESSI, OTUA, ARCELOR, CTBA, APCI AND MATERIO—OR IN A PERSONAL CAPACITY, ENABLED THE USE OF VALUABLE IMAGES AND DOCUMENTS TO ILLUSTRATE THE TEXT. IN PARTICULAR CHARLES EDELSTENNE, MATHIEU DURAND, AGNÈS LÉVITTE, MONIQUE TANGUY, FANNY BÉCUE, DANIEL GUINARD, PIERRE PARISOT, FRANÇOISE VIGIER, ANNE-MARIE BOUTIN, ÉLODIE TERNAUX, CATHERINE PROUVÉ-DERAIN, JACQUES BARSAC, PERNETTE PERRIAND-BARSAC, NADINE COLENO, AND MONIQUE DUPONT-SAGORIN.

INDEX

EDITORIAL DIRECTION: **JULIE ROUART**,
ASSISTED BY **CLÉMENT DIRIÉ**
TRANSLATED FROM THE FRENCH BY **DAVID RADZINOWICZ** (ALL TEXT EXCEPT THE FOLLOWING): **JOSEPH WEST** (FROM MATERIALS TO DESIGN) AND **MADELEINE VELGUTH** (POEM ON PAGES 34–5)
COPYEDITING: **LINDSAY PORTER**
DESIGN: **LINE MARTIN**
TYPESETTING: **THOMAS GRAVEMAKER**
PROOFREADING: **ANNE KORKEAKAVI**
COLOR SEPARATION: **DUPONT, PARIS**

DISTRIBUTED IN NORTH AMERICA BY RIZZOLI INTERNATIONAL PUBLICATIONS, INC.

SIMULTANEOUSLY PUBLISHED IN FRENCH AS
DESIGN: TECHNIQUES ET MATÉRIAUX

ENGLISH-LANGUAGE EDITION

WWW.EDITIONS.FLAMMARION.COM

06 07 08 4 3 2 1
FC0519-06-V
ISBN-10: 2-0803-0519-0
ISBN-13: 9782080305190
DÉPÔT LÉGAL: 05/2006

PRINTED IN CHINA BY IMAGO